U.S. Marine Corps Combat Water Survival

Plus

Military Training for Urban Operations

Department of Defense

Pentagon
Publishing

MCRP 3-02C

Marine Combat Water Survival

U.S. Marine Corps

PCN 144 000069 00

To Our Readers

Changes: Readers of this publication are encouraged to submit suggestions and changes that will improve it. Recommendations may be sent directly to Commanding General, Marine Corps Combat Development Command, Doctrine Division (C 42), 3300 Russell Road, Suite 318A, Quantico, VA 22134-5021 or by fax to 703-784-2917 (DSN 278-2917) or by E-mail to **morgann@mccdc.usmc.mil**. Recommendations should include the following information:

- Location of change
  - Publication number and title
  - Current page number
  - Paragraph number (if applicable)
  - Line number
  - Figure or table number (if applicable)
- Nature of change
  - Add, delete
  - Proposed new text, preferably double-spaced and typewritten
- Justification and/or source of change

Additional copies: A printed copy of this publication may be obtained from Marine Corps Logistics Base, Albany, GA 31704-5001, by following the instructions in MCBul 5600, *Marine Corps Doctrinal Publications Status*. An electronic copy may be obtained from the Doctrine Division, MCCDC, world wide web home page which is found at the following universal reference locator: **http://www.doctrine.usmc.mil**.

Unless otherwise stated, whenever the masculine gender is used, both men and women are included.

DEPARTMENT OF THE NAVY
Headquarters United States Marine Corps
Washington, D.C. 20380-1775

6 January 2003

FOREWORD

Marine Corps Reference Publication (MCRP) 3-02C, *Marine Combat Water Survival*, provides Marine Corps combat water survival techniques, procedures, and training standards. This publication also teaches Marines to cross water obstacles and perform water rescues correctly and safely.

This publication is the foundation for teaching Marines correct water survival techniques and procedures that are used throughout the Marine combat water survival program (MCWSP). Once an individual or a unit has completed the MCWSP, this publication can be used as a refresher course before water operations.

MCRP 3-02C supersedes Fleet Marine Force Reference Publication (FMFRP) 0-13, *Marine Combat Water Survival*, dated 16 September 1991.

Reviewed and approved this date.

BY DIRECTION OF THE COMMANDANT OF THE MARINE CORPS

EDWARD HANLON, JR.
Lieutenant General, U.S. Marine Corps
Commanding General
Marine Corps Combat Development Command

Publication Control Number: 144 000069 00

MCRP 3-02C Marine Combat Water Survival

Table of Contents

Chapter 1. Survival at Sea

Abandoning Ship 1-2
Jettisoning Equipment 1-2
Abandoning Ship Technique 1-3
Modified Abandoning Ship Technique 1-6
Surface Burning Oil Swim 1-9
Surviving With a Pack 1-11
Preparing Equipment 1-12
Tying Waterproof/Plastic Bags 1-13
Packing the Pack 1-13
Swimming With the Pack 1-13
Staying Afloat With a Life Preserver 1-15
Inherently Buoyant Life Preservers 1-15
Inflatable Life Preservers 1-15
Staying Afloat Without a Life Preserver 1-18
Floating With an Inflated Blouse 1-18
Floating With Inflated Trousers 1-20
Sling Method 1-20
Splash Method 1-26
Blow Method 1-32
Avoiding Heat Loss in Cold Water 1-36
Individual Protection From the Cold 1-37
Group Protection From the Cold 1-38
Drownproofing Methods 1-39
Crawl Stroke 1-40
T-Method 1-44
The Sweep 1-46

Breast Stroke....1-48
Side Stroke....1-52
Elementary Backstroke....1-57

Chapter 2. Water Rescues

Reaching Rescue Techniques....2-2
Reach....2-2
Reach From a Deck....2-3
Arm Extension....2-3
Leg Extension....2-4
Wading Assist....2-4
Throw....2-5
Lifesaving Approaches....2-7
Front Surface Approach....2-7
Rear Approach....2-7
Approach Strokes....2-8
Crawl Stroke Approach Stroke....2-8
Breast Stroke Approach Stroke....2-11
Level Offs....2-13
Front Surface Approach....2-13
Single Armpit Level Off....2-14
Double Armpit Level Off....2-16
Rescue Techniques....2-18
Wrist Tow....2-18
Single Armpit Tow....2-19
Double Armpit Tow....2-21
Collar Tow....2-22
Cross-Chest Carry....2-22
Tired Swimmer's Assist....2-24
Defense Against a Drowning Victim....2-25
Block....2-25
Wrist-Grip Escape/Wrist-Grip Escape Alternative....2-26

Front Head-Hold Escape 2-27
Rear Head-Hold Escape 2-29
Administering First Aid/Rescue Breathing 2-31

Chapter 3. Treatment of Casualties and Avoidance of Dangerous Marine Life

Drowning 3-1
Symptoms 3-1
Treatment 3-2
Hypothermia 3-2
Symptoms 3-4
Survival Time 3-4
Treatment 3-4
Heat-Related Injuries 3-6
Burns 3-7
Common Medical Problems Associated with Sea Survival 3-8
Seasickness 3-8
Saltwater Sores 3-8
Blindness/Headache 3-9
Constipation 3-9
Sunburn and Dehydration 3-9
Dangerous Marine Life 3-10
Sharks and Barracudas 3-11
Sea Snakes 3-12
Poisonous Fish 3-12
Turtles and Moray Eels 3-12
Corals 3-12
Jellyfish, Portuguese Man-of-War, Anemones, and Others 3-12
Spiny Fish, Urchins, Stingrays, and Cone Shells 3-13

Chapter 4. Negotiating Water Obstacles

Tides....................4-1
Surf....................4-2
  Plunging Waves....................4-3
  Spilling Waves....................4-4
  Surging Waves....................4-4
Currents....................4-4
  Offshore Currents....................4-4
  Rip Currents....................4-5
  Littoral Currents....................4-5
Back Bays....................4-6
Rivers and Canals....................4-6

Chapter 5. Fording Waterways

Selection of a Ford Site....................5-1
  Determine the Slope....................5-2
    Clinometer....................5-2
    Map....................5-3
    Line of Sight and Pace....................5-3
  Determine the Current Speed....................5-4
  Measure River Width....................5-5
  Calculate Downstream Drift....................5-6
The Buddy System....................5-7
Water Crossings....................5-8
  Care of Weapons....................5-8
  ISOMAT Raft....................5-9
  Poncho Raft....................5-10
  Construction of a Pack Raft....................5-12
  Single-Rope Bridge....................5-13
  High and Dry Crossings....................5-17

Swift Current Crossings 5-18
Slow Current Crossings 5-19
Removal 5-19

Appendix. Knot Tying

Square Knot Appendix-1
Bowline Appendix-2
Hitches Appendix-2
Half Hitch Appendix-2
Two Half Hitches Appendix-3
Round Turn and Two Half Hitches Appendix-3
Butterfly Knot Appendix-4

Throughout history, water has posed special challenges to Marines and Sailors during times of both peace and war. Therefore, the inherent dangers associated with waterborne operations demand that Marines and Sailors receive proper water survival training. Combat units that are confident in their ability to work in and around water can use the water to their advantage in combat, and history is filled with examples of how the proper preparation or training for survival in water reduced or averted disaster.

USS *Indianapolis*

On Sunday, 29 July 29 1945, the heavy cruiser USS *Indianapolis* (the *Indy*) was en route to the Philippine Sea. Shortly before midnight (about 39 hours out of port), the *Indy*, running blacked out and unescorted, was rocked by two explosions on her starboard side. With communications smashed, the ship could not signal its distress and sank within 15 minutes.

Three life rafts and a floater net supported a few survivors, but the rest of the survivors drifted about, held up by rubber life belts or Mae Wests. About 60 seamen died the first night. Survivors assumed the ship would be reported overdue in Leyte, and that they would be rescued within 2 days. By Monday evening, however, panic began to set in as some life jackets lost their buoyancy from the long immersion. Some men even fought over life jackets, which resulted in at least 25 deaths. No one dared sleep for fear of losing his jacket.

Throughout the next several days, in-transit aircraft flew nearby without spotting the desperate seamen. As best they could, the men kept together, some tying long ropes to each other, floating like corks on a net.

Not until late Thursday morning, 3 1/2 days after the ship sank, were the men discovered. Luckily, a plane on a routine flight over the area spotted the survivors. When ships picked them up that night, the survivors learned they had never been reported overdue. Every one of the 1,196-man crew was a casualty; 880 crewmembers were listed as dead or missing. Although many lives were lost, the innovative and expedient use of flotation devices and float techniques employed by the survivors helped save hundreds of lives.

USS *America*

On Thanksgiving eve, 23 November 1995, the USS *America* made its way through the Arabian Sea. Twenty-year-old, Marine Lance Corporal (LCpl) Zachary Mayo was unable to sleep and, wanting some fresh air, made his way onto an open-air platform near the aircraft hangar bay, which was three levels below the sleeping quarters. While he was on the platform, the ship veered suddenly, throwing LCpl Mayo through the platform's protective bars and into the sea, 30 feet below.

Frantic, LCpl Mayo called out in vain to the watchmen on the flight deck, which was 64 feet above him. It soon became clear to him that the USS *America* would keep its course into the Gulf of Oman until his absence was discovered at morning muster. The LCpl took a moment to consider his situation. Since land was at least 100 miles away, swimming was suicide; he would have to stay afloat until a search party found him.

Using the techniques he had learned during combat water survival training, LCpl Mayo made a flotation device out of his coveralls and tried to relax. Meanwhile, business continued as usual aboard the USS *America*. Since LCpl Mayo was on special assignment with the hazardous materials division, his absence went unnoticed until a petty officer asked several Sailors if they had seen their shipmate recently. By the time a roll call had been completed, LCpl Mayo had been adrift at sea for over 24 hours. Although three, fixed-wing Viking aircraft were deployed to search for LCpl Mayo, most people aboard ship feared the worst.

After 34 hours at sea, LCpl Mayo was discovered by fishermen on a Pakistani fishing boat. LCpl Mayo's survival is a testament not only to his incredible physical courage, but also to the soundness of the lifesaving training and techniques he received during combat water survival training.

Chapter 1
Survival at Sea

As a Marine, you face a variety of potential water emergencies whenever you cross expanses of water: ships, watercraft, and amphibious assault vehicles (AAVs) can sink; aircraft can crash into the sea; or you can accidentally fall into the water. However, there are some basic precautionary measures you can take to protect your safety and reduce your chances of becoming a water casualty. Determine the following information as soon as you board any type of vessel. Your knowing the following information may save your life or the life of your fellow shipmates.

- How many life preservers and lifeboats/rafts are on board?
- Where are the life preservers and lifeboats/rafts located?
- What type of unit survival equipment is on board?
- Are individual survival kits issued to each person on board?
- How much food, water, and medicine do the survival kits contain? When was the last time the contents were inspected for proper quantities and shelf life expiration?
- Is there sufficient survival equipment available for the number of personnel?
- How many other personnel are there on board, and where are they located?
- What are the egress procedures for the ship, boat, watercraft, AAV, or aircraft?

Abandoning Ship

When you embark on a Navy ship, you will receive abandoning ship instructions from Navy personnel. If given the order to abandon ship, report to your designated assembly area and put on a life preserver. DO NOT inflate the life preserver until you are clear of the ship. Torn life preservers will not inflate and inflated life preservers can block you, and those behind you, from exiting the ship. A flotation device that has been inflated may also burst if you jump from a significant height. See pages 1-15 through 1-35 for staying afloat with and without a life preserver.

DO NOT remove your clothing, boots, or shoes before abandoning ship. Your trousers and blouse may be the only flotation devices available if your life preserver is faulty or becomes damaged, and your clothes can provide some insulation from the cold water. However, remove your soft cover and place it in a cargo pocket for later use. The soft cover is both lightweight and a good protection against sunburn caused by the sun's rays reflecting off the water.

Jettisoning Equipment

Equipment should be kept properly packed and waterproofed in case you have to abandon ship. If entering the water from a height greater than 30 feet, wearing your equipment (e.g., pack, helmet, gas mask) could cause injury. Upon impact with the water, the helmet will "cup" air inside of it. The chin strap may also create a "hanging effect" as you submerge from the force of the fall. Therefore, you should remove your helmet and gas mask before abandoning ship.

If you are unable to maintain buoyancy due to the amount of equipment secured to your pack and body, then jettisoning some of your equipment may become necessary. Equipment that you should always retain includes canteens of freshwater, first aid kit, soft cover, and survival kit. The survival kit should include first aid items, water purification tablets or drops, fire starting equipment, signalling items (e.g., flashlight, strobe light, chemlights), food procurement items, and shelter items. Other items in a survival kit include sunburn lotion and lip balm, knife, goggles/sunglasses, plastic bag, matches and lighter, mirror. See Marine Corps Reference Publication (MCRP) 3-02F, *Survival*, for a detailed list of survival items and applications.

Abandoning Ship Technique

When abandoning ship, safety considerations must be observed. Use the following technique when abandoning ship without your combat gear:

Place your hands on their opposite shoulders, forming a criss-cross pattern.

Step to the edge of the ship's deck and check the water below for debris or survivors. If the water is clear, look straight ahead and prepare to jump. If the water is not clear, move to another location.

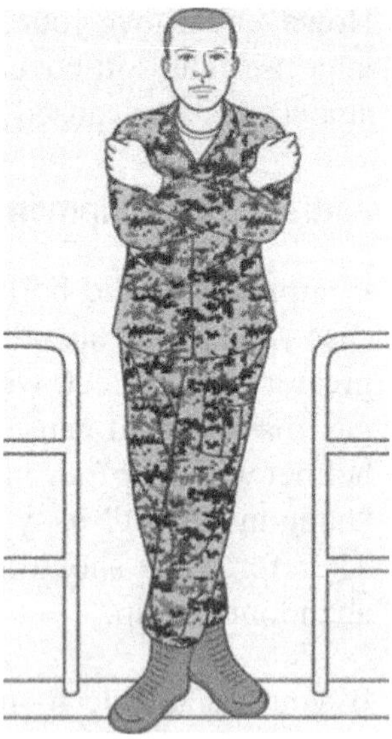

NOTE: DO NOT hold your nose as you abandon ship. If you do hold your nose, the force of impact into the water could jar your arm and hand and cause you to break your nose.

Step off the side of the ship with a smooth, 30-inch stride. DO NOT DIVE OFF THE SHIP. DO NOT LOOK DOWN AT THE WATER. LOOK STRAIGHT AHEAD. Looking down at the water can render you unconscious or cause injuries upon impact.

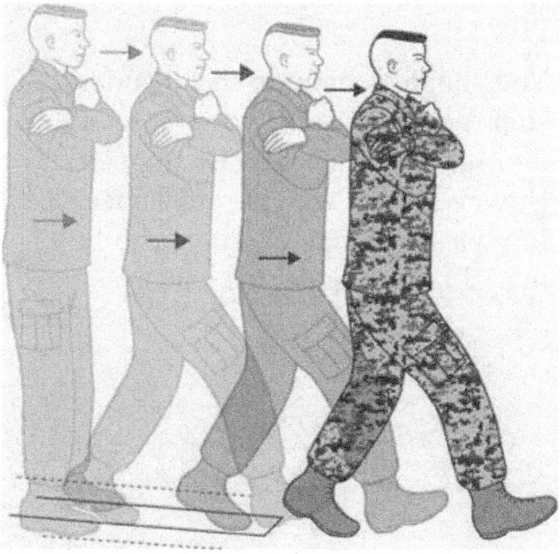

Bring your trailing leg forward during the fall. Cross your trailing leg behind your leading leg.

Keep your head parallel to the water's surface until hitting the water.

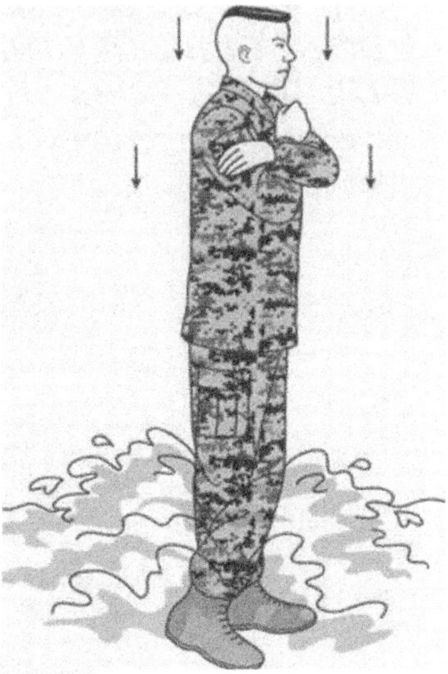

You should remain in the abandon ship position until your descent into the water has almost stopped. However, the weight imbalances in your body may cause you to be in a "J" shape under the water

.

Once your downward motion has ceased, your feet may be parallel with the ocean bottom or you may be nearly inverted with your feet over your head. To counteract potential disorientation, you should pause briefly and allow the natural buoyancy of your torso to bring your body to a nearly upright position.

Floating debris can cause hazards. Therefore, you should swim upward, extending one arm (hand is shaped as a fist) upward to feel for obstructions. If you encounter debris, try to push it away or surface in a different location.

Swim away from the ship. DO NOT LOOK BACK AT THE SHIP. Looking back could slow your movement away from the area. Remember, your objective is to leave the area as quickly as possible because—

- Equipment and debris may be falling from or spilling out of the ship.
- Additional casualties can occur if individuals abandoning the ship fall on top of swimmers already in the water.
- Swimmers close to the sinking ship may get pulled underneath the water by the suctioning effect of the ship as it goes under.

Modified Abandoning Ship Technique

When abandoning ship while wearing full combat gear (weapon, helmet, and a properly waterproofed pack), safety considerations must be observed. The modified abandoning ship technique is used while wearing full combat gear and exiting from a height less than 30 feet.

WARNING

When exiting from a height greater than 30 feet, remove your helmet and pack. Fasten the helmet to the pack or place it inside the pack before jettisoning. If jettisoning gear from the ship or aircraft, check the water below for survivors before throwing the gear forward of the intended jump area. Once in the water, retrieve your gear and swim out of the area.

Place your weapon over one shoulder, muzzle down, with the weapon parallel to your side.

Place your arm and hand along the weapon and hold it to your side.

Take your free hand and place it on top of your helmet to prevent neck/spinal injury from the force of the water pulling upward on the helmet as your body enters the water.

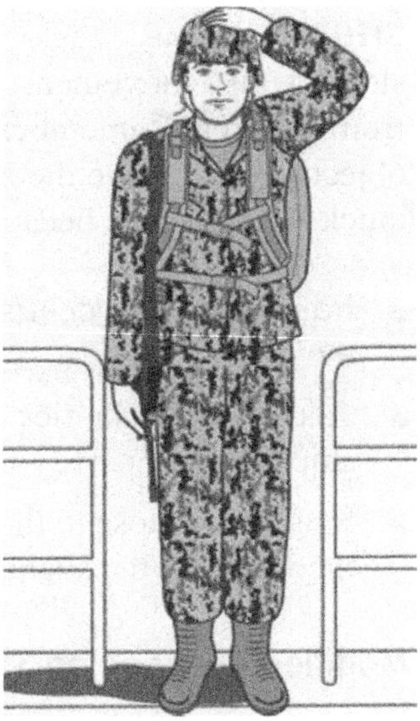

NOTE: An alternate method is to place your weapon over one shoulder, muzzle down, with the weapon parallel to your side. Reach across your body and grasp the sling of your weapon and hold it to your body. Take your free hand and place it on top of your helmet.

Step to the edge of the platform and check the water below for debris and survivors. If the water is clear, look straight ahead and prepare to jump. If the water is not clear, move to another location.

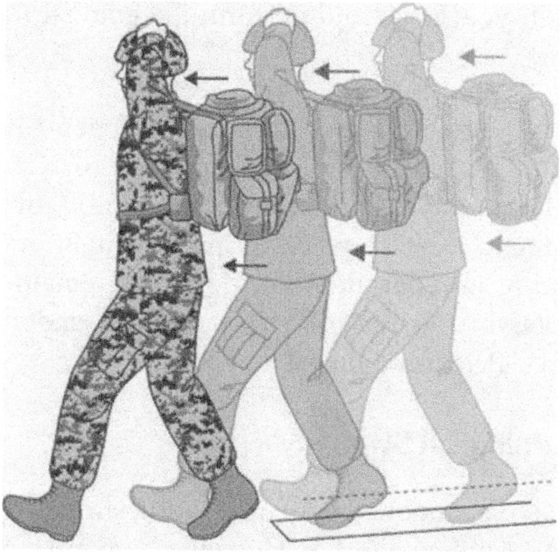

Step off the side of the platform with a smooth, 30-inch stride. DO NOT DIVE OFF THE SHIP, DO NOT LOOK DOWN AT THE WATER, LOOK STRAIGHT AHEAD. Looking down at the water can render you unconscious or cause injury upon impact.

Bring your trailing leg forward during the fall. Cross your trailing leg behind your leading leg.

Keep your head parallel to the water's surface until hitting the water.

You should remain in the modified abandoning ship position until your descent into the water has almost stopped. The buoyancy of a properly waterproofed pack will immediately pull you to the surface. Once you break the water's surface, unsling your weapon and loop the sling over your head. You may have to seesaw the sling so as to ensure the sling passes between your helmet and pack. Once over your head, loop the sling around your neck with the weapon aligned in the center of your body, muzzle down. Lean back on your pack and perform the combat travel stroke to exit the area.

> *NOTE:* You may also remove the pack (unsnap the quick-release strap on one side of the pack, and pass the strap to your other hand in order to maintain contact with the pack, which is functioning as your flotation device). Remain horizontal in the water with your pack under your chest. Perform a breast stroke kick.

Surface Burning Oil Swim

After you have abandoned ship, rise to the surface using the techniques shown on the page 1-6. However, you must remember that fuel from sinking ships or downed aircraft will float on the surface of the water. Therefore, you must move clear of the floating fuel by swimming away from the ship or aircraft as soon as possible. Either swim upwind (into the wind) of the ship/aircraft or swim against the current. Either method allows you to move away from the fuel and the wind/current will push the fuel past you. To properly execute a surface burning oil swim—

Extend your arms overhead as far as possible.

Wave your arms back and forth vigorously to splash a hole while moving upward.

Splash as long as possible to push burning fuel away from the surfacing area.

Use your arms and hands to sweep away fuel and debris.

Kick your legs in a constant breast stroke kick.

Extend your arms (palms outward) forward on the surface, arms shoulder-width apart.

Pull your hands in and back toward the chest.

Stop your hands in front of your face and rotate them so that your palms face forward (roughly halfway out of the water).

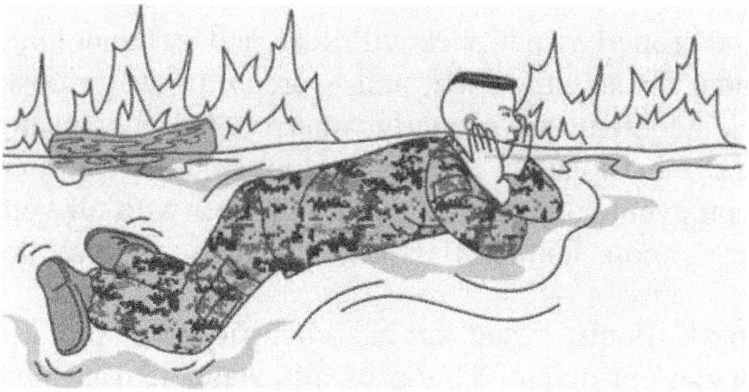

Sweep your arms forward to a full extension at the shoulder width. This splashes debris, oil, or burning liquids aside. To reduce the chance of fatigue, use two short splashes to the front to extend the path.

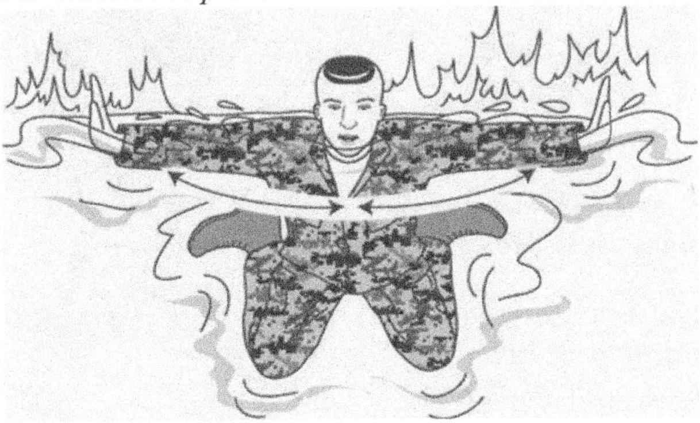

Repeat the preceding step as necessary while swimming clear of the area.

Surviving With a Pack

If packed properly, your pack will float, and it is your key piece of equipment for staying afloat and overcoming water obstacles. If the pack's contents are properly waterproofed, it can support you (with a combat load) in the water. Buoyed up by a waterproofed pack, you eventually emerge from the water with all your equipment (e.g., boots, helmet, flak jacket, weapon, survival items).

Your pack floats based on a scientific principle known as Archimedes' principle. This principle states that an object submerged in a liquid is buoyed up by a force equal to the weight of the liquid displaced (pushed aside) by the object. If the weight of the displaced liquid is greater than the weight of the object, the object floats. If the weight of the displaced liquid is less than the weight of the object, the object sinks. For example, a machine gun sinks in the water, but it still weighs less in the water than it does on land. Even though a machine gun sinks, it is still buoyed up by a force equal to the weight of the water it displaces. For this reason, you should not try to hold yourself or your equipment any higher out of the water than they would naturally float; doing so wastes both energy and body heat.

Preparing Equipment

Before packing your pack, you must prepare your gear/equipment. Tape or pad all sharp edges and equipment corners. Ideally, your gear/equipment is placed in plastic bags and the plastic bag is then placed inside the standard-issued, rubberized, waterproof bag. Waterproof bags are not completely water tight; henceforth, the added protection of first wrapping the gear/equipment in a plastic bag, then placing the plastic bag inside the waterproof bag. Large plastic bags (e.g., trash bags) work well for bulky equipment (e.g., sleeping bags, field jackets, shelter halves, gas masks). Small plastic bags work best for small items (e.g., shaving gear).

> *Note:* If a gas mask must be carried outside the pack, cover it with a waterproof bag.

You will need the following items, which are available through the supply system, to prepare your equipment for packing:

- Waterproof bags:
  Pistol bag, plastic size #1 (8" x 18").
  Rifle bag, plastic size #2 (10" x 56").
  Machine gun bag, plastic size #3 (15" x 56").
  Multipurpose bag, cover, plastic size #4 (20" x 84").

- Panama work vest (life preserver).
- Small plastic bag, self sealing (6" x 6" and 12" x 12").
- Riggers tape, olive drab (2 1/2").
- Willie peter bag, waterproof clothing.

Tying Waterproof/Plastic Bags

Try to remove the excess air from the waterproof/plastic bag before securing the bag's opening. If filled with air, the bag can burst if pressed from the outside. Perform the following steps to tie a waterproof/plastic bag:

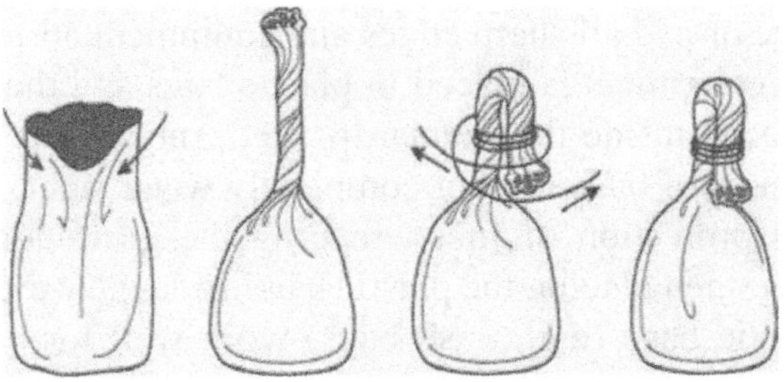

Packing the Pack

Place filled and tied bags inside the pack. Carefully handle sharp items (e.g., tent pegs, poles) to prevent puncturing the bag. Place items in the pack in order of expected use. Close the pack and its compartments. Attach sleeping mats or bags as high as possible on the outside of the pack.

Swimming With the Pack

A useful technique for propelling yourself forward while wearing a pack is the combat travel stroke. This technique allows you to float nearly horizontal and to propel yourself forward with bicycle-style kicks and breast stroke-style arm sweeps. To execute the combat travel stroke—

Body position
Keep the upper part of your body prone to the water, let your legs dangle horizontally, and keep your face up.

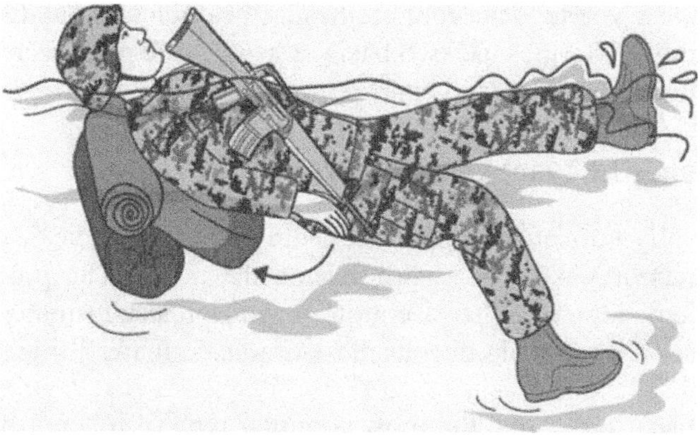

Arm action
Extend your hands out in front of your waist. Sweep your arms slightly downward and back 90 degrees to propel your body through the water. Move your hands back to the front of your waist. Repeat.

Leg action
Continuously move your legs in a bicycle-like movement, bringing your knees up high and step out.

Breathing
Keep your face out of the water during the stroke, and breathe freely.

Staying Afloat With a Life Preserver

The best form of flotation is to find any kind of floating object that will keep you and your equipment out of the water or minimize your exposure to the water. Life preservers are the best method, as they allow you to wear your clothes for heat retention and sunburn prevention. Marines use two basic classes of life preservers: inherently buoyant life preservers and inflatable life preservers.

Inherently Buoyant Life Preservers

Inherently buoyant life preservers are either vest-type (worn like a jacket) or yoke-type (worn around the neck). The preserver's outer envelope is either a cotton or water resistant material that encloses a removable fibrous glass or plastic foam filling.

In the Marine Corps, the most common type of inherently buoyant life preserver is the vest-type with collar, known as the kapock preserver. The kapock consists of collar straps, upper front chest straps, leg straps, and waist drawstrings that secure the preserver to you. The leg straps, which are fitted on both sides of the life preserver, ensure that the preserver remains around your chest while you are in the water. A chest strap is attached to the life preserver to facilitate lifting you out of the water. The strap can also be attached to other survivors or to lifeboats to reduce the fatigue that results from holding onto a floating/secured object by hand.

Inflatable Life Preservers

Marine Corps aircraft and AAVs have inflatable life preservers on board. The inflatable life preserver issued to Marines is known as the LPP (life preserver personal). The LPP is capable of both oral inflation and CO_2 cartridge inflation. The LPP consists of buoyancy chambers, CO_2 inflator, and an oral inflation tube. The buoyancy chamber is deck gray in color and is made from a neoprene-coated nylon fabric.

Inflatable life preservers must be stored in a cool, dry place. Heat, moisture, and light cause deterioration of the life preserver material. Do not stow CO_2 cylinders near steam lines or radiators. Heat can increase the pressure inside the cylinders causing them to explode. Avoid sharp edges in stowage. Sharp edges increase wear and tear on the life preservers and may also puncture inflatable buoyancy chambers.

> **CAUTION:** Do not inflate the life preserver until you are clear of the aircraft, ship, or vehicle. Torn life preservers will not inflate and inflated life preservers can block you, and those behind you, from exiting the aircraft, ship, or vehicle.

To don and adjust the LPP—

Remove the LPP from its storage container.

Fasten the belt fasteners in front with the pouch in the rear.

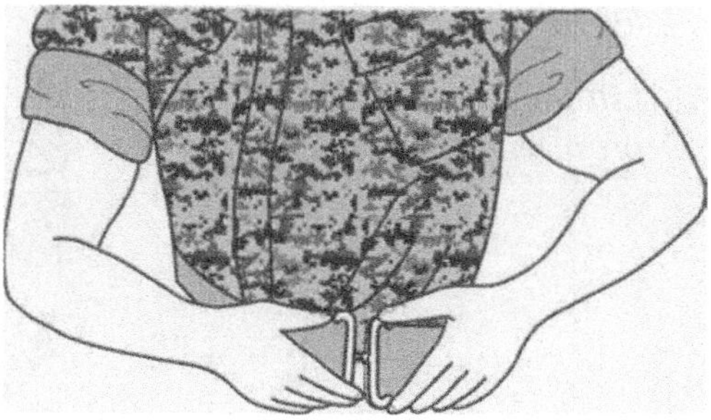

Adjust the belt to fit; secure any excess belt by mating the hook and pile tape.

Rotate the pouch to the front and re-adjust the belt if necessary.

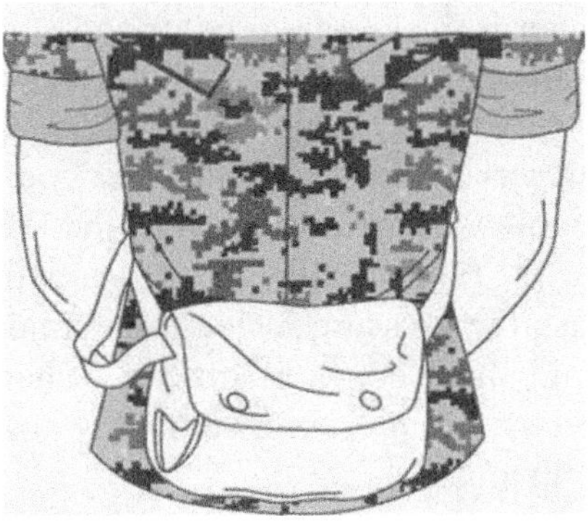

Open the snap fasteners on the pouch and unfold the life preserver.

Place the deflated life preserver over your head.

Place the storage container into the pouch after donning the life preserver.

Lift the lower end of the life preserver out of the pouch.

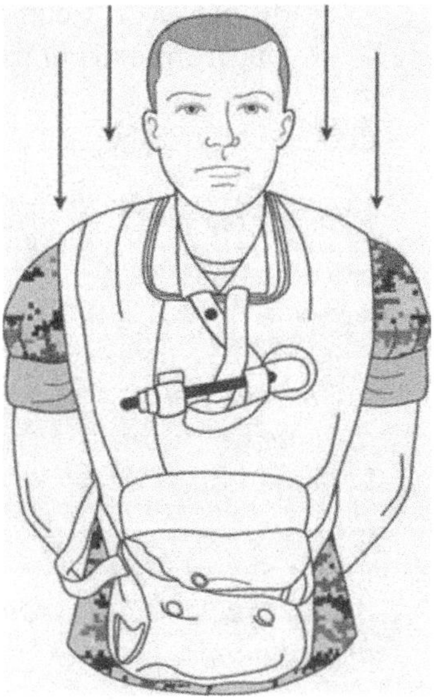

Inflate the life preserver by pulling on the lanyard attached to the CO_2 inflation valve or by blowing on the end of the oral inflation valve.

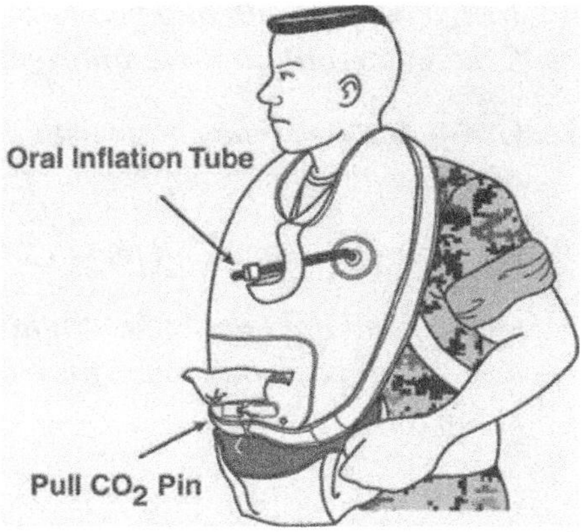

Staying Afloat Without a Life Preserver

You may be in open water without any floating objects or a life preserver to help you survive. If so, uniform blouses and trousers can be made into expedient flotation devices.

Floating With an Inflated Blouse

It is possible to float by a bubble of air trapped in the shoulders of your blouse. The air rises to the back and shoulders of the blouse and supports you at the water's surface. An inflated blouse is also a temporary flotation device used by weaker swimmers while trying to remove their trousers. There is a primary and an alternate way to create a bubble of trapped air in a blouse—

Primary Method
Turn the collar inside the blouse to help create a seal.

Unbutton top button and pull collar around mouth and nose.

Take a deep breath and bend forward slightly at the waist. Exhale one-half to three-quarters of a breath into the blouse.

Grasp and twist the collar with one hand to create a seal, this prevents air from escaping out from the collar.

Use your free hand and feet to stroke and kick to the surface.

Gather and hold the blouse tightly at the collar and stomach level to prevent the blouse from losing air if it floats up too high.

Splash water on the blouse periodically to prevent the material from drying, dry material allows air to escape.

Repeat inflation as required.

Alternate Method
Turn the collar inside the blouse to help create a seal.

Unbutton the second button from the top.

Take a deep breath and bend forward slightly at the waist.

Place your mouth and nose inside the hole created by the open button and exhale one-half to three-quarters of a breath into the blouse.

Grasp material at the unbutton portion and pull downward.

Use your free hand and feet to stroke and kick to the surface.

Splash water on the blouse periodically to prevent the material from drying, dry material allows air to escape.

Repeat inflation as required.

Floating With Inflated Trousers

In warm water, trousers can be used as a primary expedient flotation device. However, in cold water, submerging your head to remove and inflate your trousers results in heat and energy losses that negate the benefit of using the trousers as a flotation device. Once your trousers are inflated, you float motionlessly as if wearing a life preserver. If needed, assume the heat escape lessening posture (known as the HELP position, see page 1-37) to slow heat loss. As trousers dry, air leaks out of the legs. To slow this process, occasionally splash water on the fabric. Reinflate trousers as needed.

Sling Method

The sling method works if you are a strong swimmer or naturally very buoyant. Take the following steps to inflate trousers using the sling method:

Take a deep breath, bend over, and remove your boots.

NOTE. Retain your boots. Tie the boot laces together and suspend the boots from your blouse or hang them around your neck so that they rest on your chest.

Remove your trousers. Button or zip the trouser's fly closed. This allows you to control airflow.

Tie the bottoms of the trouser legs in a square knot. (The appendix illustrates various knots.)

Ensure that the front (fly) of the trousers faces you.

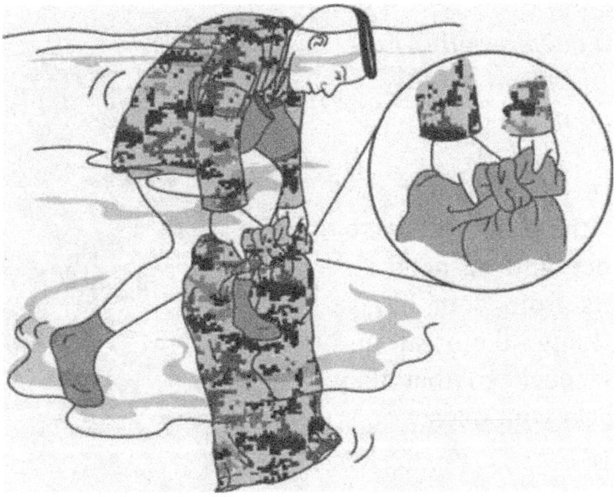

Hold the trousers above the water's surface and behind your head. Grasp both sides of the waistband and open with both hands.

Kick strongly to stay on top of the water while slinging the trousers overhead in order to trap air into them.

Once the waistband is submerged in the water, air is trapped in the legs.

Hold and seal the waistband underwater.

Slip the inflated legs over your head. Hold the waistband in toward your chest, the fly facing your body. To prevent air from escaping from the trousers, seal the waistband by either folding it or twisting it.

Lie back and relax, resting the back of your neck against the knot.

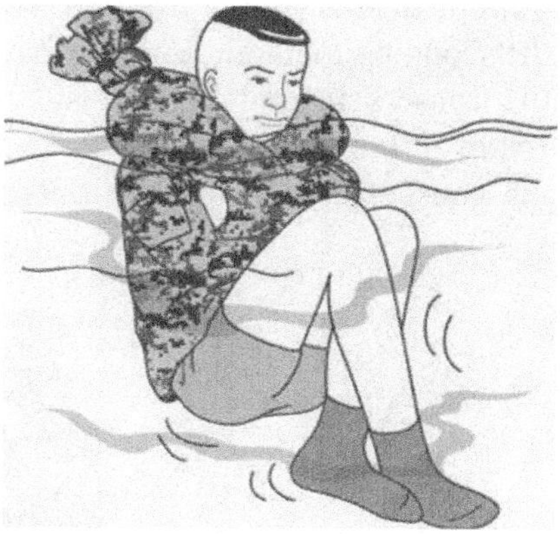

Splash water on the trousers periodically to prevent the material from drying. Dry material allows air to escape.

To replenish air in the trousers, you will use a technique known as the scooping method. With one hand on the open waistband, extend the trousers in front of you just below the surface of the water and scoop air bubbles with your free hand into the open waistband until the trousers have sufficient air. Repeat as necessary.

Splash Method

The splash method is an alternative to the sling method. As with the sling method, you must kick strongly to remain at the surface. To inflate trousers using the splash method, perform the following:

Take a deep breath, bend over, and remove your boots.

NOTE: Retain your boots. Tie the boot laces together and suspend the boots from your blouse or hang them around your neck so that they rest on your chest.

Remove your trousers. Button or zip the trouser fly closed. This allows you to control airflow.

Tie the bottoms of the trouser legs in a square knot. (The appendix illustrates various knots.)

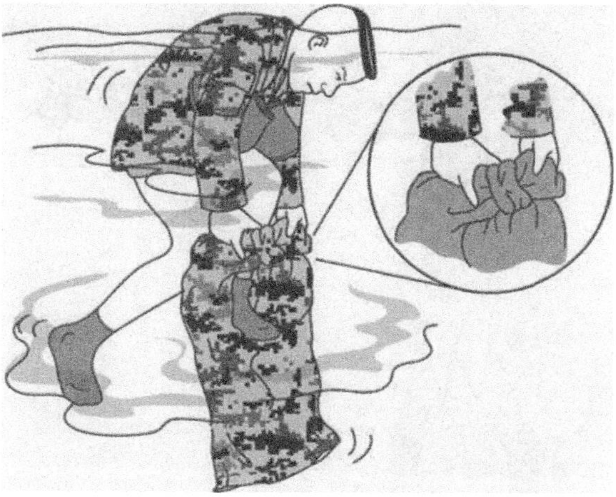

Ensure that the front (fly) of the trousers faces you.

Hold the trousers at the water's surface out in front of you by the waistband with the fly up.

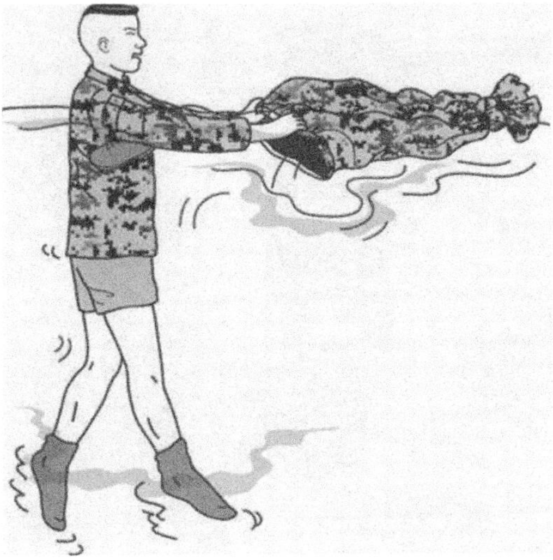

Grasp the waistband at the surface with one hand. Insert your free hand into the waistband, palm down.

Flutter your hand rapidly to create bubbles. This sends a mixture of water and air bubbles into the trousers. The water passes through the fabric. The air remains trapped in the legs.

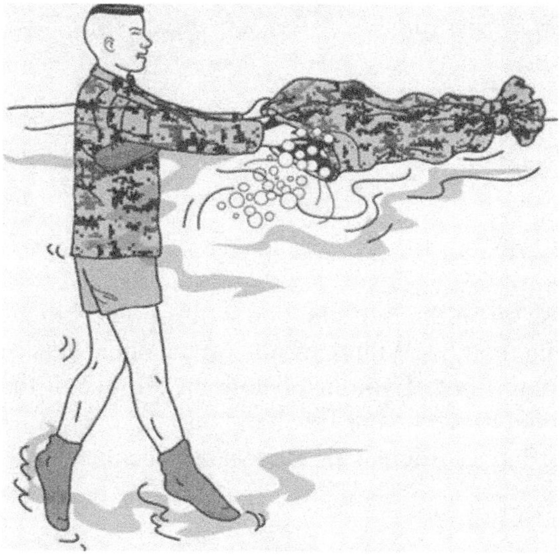

Hold and seal the waistband underwater.

Slip the inflated legs over your head. Hold the waistband in toward your chest, the fly facing your body. To prevent air from escaping from the trousers, seal the waistband by either folding it or twisting it.

Lie back and relax, resting the back of your neck against the knot.

Splash water on the trousers periodically to prevent the material from drying. Dry material allows air to escape.

To replenish air in the trousers, you will use a technique known as the scooping method. With one hand on the open waistband, extend the trousers in front of you just below the surface of the water and scoop air bubbles with your free hand into the open waistband until the trousers have sufficient air. Repeat as necessary.

Blow Method

The blow method is an alternative to the sling method. Use the blow method if you are a weak swimmer. Take the following steps to inflate trousers using the blow method:

Take a deep breath, bend over, and remove your boots.

NOTE: Retain your boots. Tie the boot laces together and suspend the boots from your blouse or hang them around your neck so that they rest on your chest.

Remove your trousers. Button or zip the trouser fly closed. This allows you to control airflow.

Tie the bottoms of the trouser legs in a square knot. (The appendix illustrates various knots.)

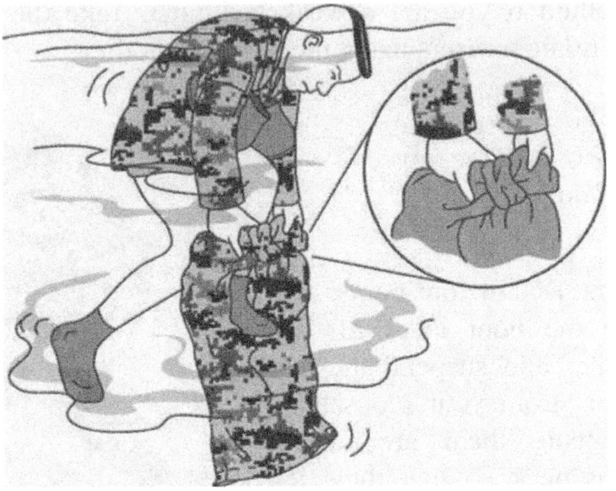

Ensure that the front (fly) of the trousers faces you.

Hold the trousers at the water's surface. Grasp both sides of the waistband and open with both hands.

Take a deep breath.

Drop 2 feet below the water's surface, pulling the waistband underwater.

Hold the waistband open with both hands and blow air into the trousers.

To fill the trousers with air, surface while keeping the waistband underwater, breathe in again, drop below the water's surface, and blow air into the trousers. Repeat these steps until the trousers are filled sufficiently. Once trousers are filled—

Hold the waistband underwater. Twist and pinch it off.

Slip the inflated legs over your head. Hold the waistband in toward your chest, the fly facing your body. To prevent air from escaping from the trousers, seal the waistband by either folding it or twisting it.

Lie back and relax, resting the back of your neck against the knot.

Splash water on the trousers periodically, to prevent the material from drying. Dry material allows air to escape.

To replenish air in the trousers, you will use a technique referred to as the scooping method. With one hand on the open waistband, extend the trousers in front of you just below the surface of the water and scoop air bubbles with your free hand into the open waistband until the trousers have sufficient air. Repeat as necessary.

Avoiding Heat Loss in Cold Water

The rate of heat exchange in the water is about 25 times greater than it is in air of the same temperature. When you are immersed in cold water, hypothermia occurs rapidly due to the decreased insulating quality of wet clothing and as a result of water displacing the layer of still air that normally surrounds the body. You also lose about 50 percent of your body heat through your head; therefore, keep your head out of the water. Other areas of high heat loss are the neck, the armpits/sides, and the groin.

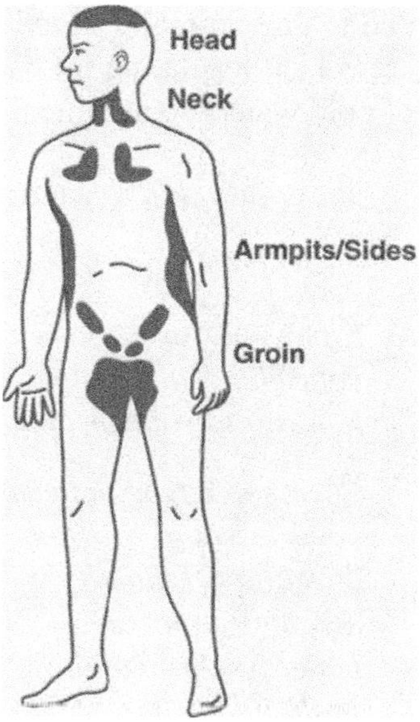

In cold water, DO NOT SWIM TO STAY WARM. Swimming, even with a slow and steady stroke, produces a lot of heat that is lost in the water. The heat loss can produce hypothermia that slows body functions and can result in serious injury or death. Remaining motionless conserves body heat three times longer than swimming. SWIM only if you have a flotation device and the shoreline is visible.

Individual Protection From the Cold

If you are equipped with a life preserver, assume the heat escape lessening posture (known as the HELP position) to slow heat loss and to protect major blood vessels near the body's surface. These areas lack insulating fat and are vulnerable to the chilling effects of cold water. To assume the HELP position—

Tuck your chin down tightly to cover your throat.

Draw your legs up in a fetal position to protect the groin.

Place your arms across your chest, tuck your hands into your armpits.

Wear some type of head covering (e.g., stowed cover, towel, handkerchief) to lessen heat loss through the scalp if head covering is available.

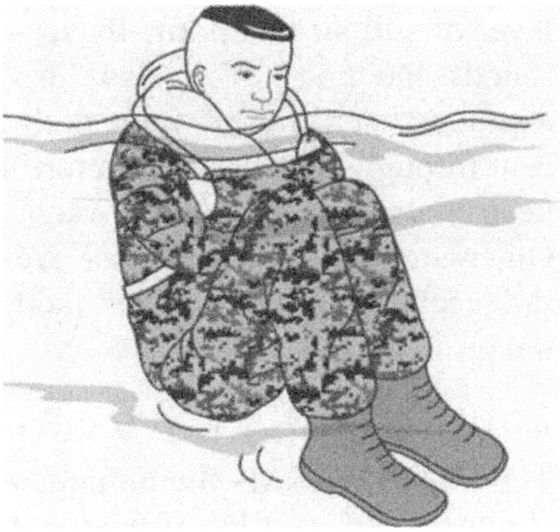

Group Protection From the Cold

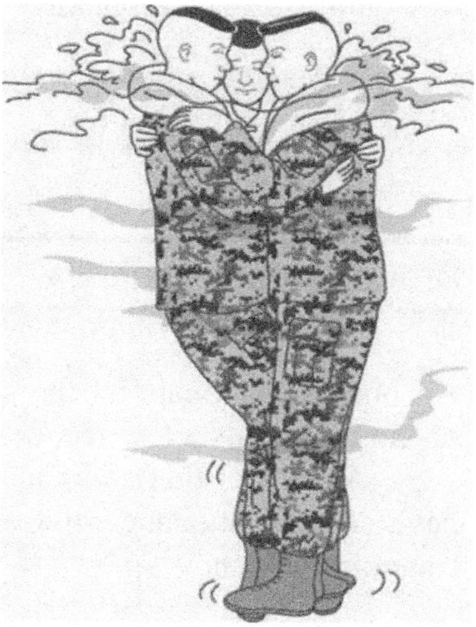

If three or more Marines are in the water and are equipped with life preservers, they should wedge tightly together and lock arms to form a circle known as a huddle position. This position protects vulnerable areas from heat loss.

A casualty who is suffering from the effects of the cold can be placed within the huddle to be surrounded by warmer water. If in the water for a prolonged period, it is recommended that everyone be rotated inside the huddle to maintain or re-warm each person's internal core temperature. If there are more than five Marines, they should make clusters of huddle positions.

Contact with other swimmers provides survival advantages—

Creates a larger target for search and rescue aircraft.

Provides additional warmth in cold water.

Improves morale.

Re-establishes the chain of command.

Reduces shock and panic.

Provides opportunities to administer first aid.

Supports exhausted Marines.

Drownproofing Methods

An object that floats on the surface has positive buoyancy. An object that floats a few feet beneath the surface has neutral buoyancy. An object that sinks has negative buoyancy. Most people have positive buoyancy and will float at the water's surface. To test your buoyancy—

Stand in water that is at your chest level or deeper.

Take a full breath.

Bend forward slowly.

Relax and wait.

If you have positive buoyancy, you will slowly rise to the surface. If you have neutral buoyancy, you will float a few feet beneath the surface. If you have negative buoyancy, you will sink. Regardless of whether you naturally float or sink, you can remain at the surface for extended periods without a life preserver if you exercise the appropriate drownproofing methods or survival strokes, which are based on your buoyancy.

Drownproofing methods consist of the T-method and the sweep. The breast stroke, side stroke, and elementary backstroke are the most common survival strokes. With any drownproofing method or survival stroke, remember the acronym **SAFE**:

SLOW EASY MOVEMENTS
Move slowly to conserve energy and minimize heat loss.

APPLY NATURAL BUOYANCY
Use natural buoyancy to support the body.

FULL LUNG INFLATION
Fill the lungs with each breath. Do not hold air in the cheeks.

EXTREME RELAXATION
Tight muscles are denser than relaxed ones and do not float as well.

Crawl Stroke

The crawl stroke, sometimes called the front crawl or free style, is the fastest stroke. To execute the crawl stroke—

Body position
Lie horizontal, on your stomach, in the water.

Look forward and downward at a 45 degree angle with the waterline between your eyebrows and hairline. Your head position is important as it assists in cutting a path through the

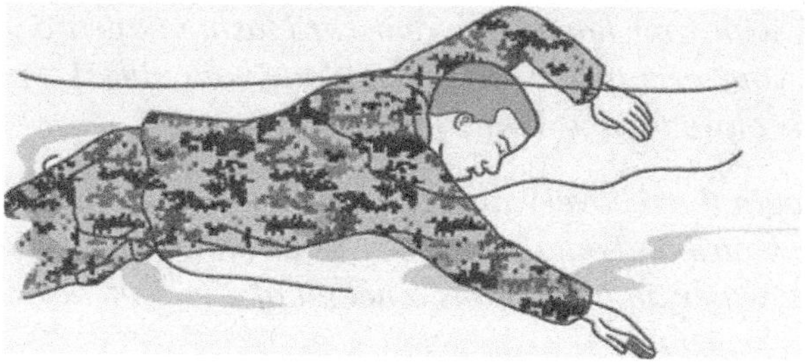

water. If your head is too high in the water, your lower body sinks significantly, making your stroke less efficient. If your head is too low in the water, water washes over your shoulders and neck, causing unnecessary drag.

Arm action
Arm action occurs in three phases: catch, propulsion, and recovery.

Fully extend one arm forward of your body, this positions your hand to catch the water in preparation for the propulsion phase.

To catch the water, bend your wrist (with your palm pointing outboard) and make an "S" shape (or inverted "S" shape) with your hand, ensuring that your hand does not cross the center of your body. Your left hand makes an "S" shape and your right hand makes an inverted "S" shape with your hand finishing at shoulder level.

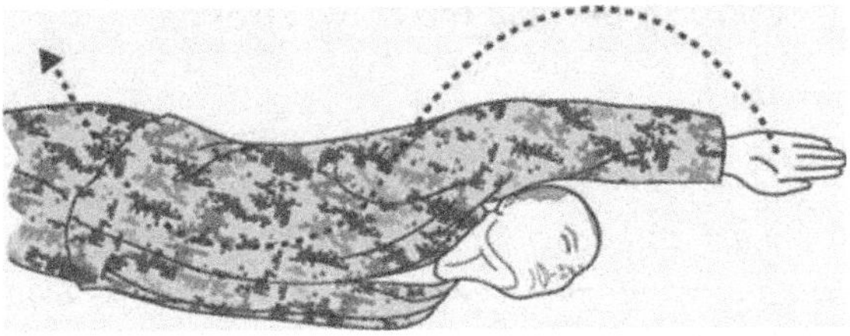

Push with your hands in a rearward fashion toward your feet until your arm is fully extended along your side, keeping your hands close to your body.

To begin the recovery phase, bend your arm at the elbow and raise your hand out of the water. Your hand breaks the surface of the water and maintains a height of 2 to 3 inches above the water's surface.

With your hand and arm moving in a forward manner, bring your hand past your head until your arm is about three-quarters of the way extended. At this point, turn your hand with your palm outboard so as to allow your thumb and fore-finger to enter the water first.

Once your hand enters the water, continue to push your arm forward until it is fully extended.

You are now prepared to catch the water again. These steps are performed in an alternating pattern: when one arm is catching and propelling, the other arm is recovering.

Leg action
Use a flutter kick to create the leg action for the crawl stroke. This kick is used for both propulsion and keeping the lower body horizontal with the water's surface. The flutter kick is an alternating leg action: one leg is kicking in a downward motion while the other leg is recovering to the surface to prepare for the next kick. The size of the kick ranges from 12 to 15 inches and depends on your height.

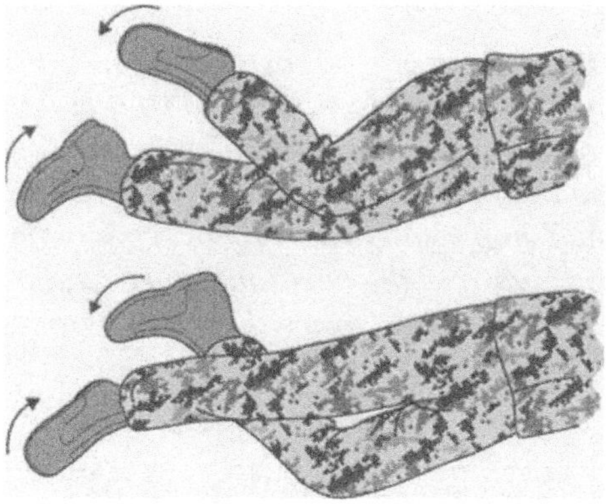

Maintain your legs in a semi-rigid manner.

Generate power for the kick from the hips.

Keep your feet loose at the ankles, they trail behind your legs and act as "flippers."

Execute the propulsion phase of this kick with a downward thrust of your leg.

Execute the recovery phase by pushing your leg back to the surface. This phase is complete when your foot reaches the surface.

Breathing
Breathe during the recovery phase of the arm action.

Roll your body and rotate your head to the side of your body where the arm recovery is occurring, this rolls the water away from your mouth. Keep your chin pushed back toward your shoulder.

Exhale while your face is still submerged and inhale when your face breaks the surface.

Breathe either bilaterally or rhythmically. To breathe rhythmically, breathe on the same side of your body every time your arm cycle occurs. Bilateral breathing requires you to breathe every one and a half arm cycles. To breathe bilaterally, breathe when your right arm is recovering, your face goes back in the water and the next time you breathe is when your left arm is starting its recovery three arm strokes later.

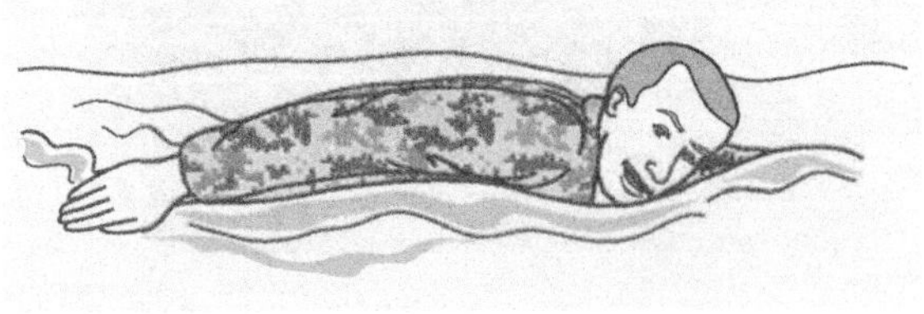

Bilateral breathing is the preferred method for breathing; it prevents the chance of hyperventilation and allows your body to maintain a lateral position in the direction you are swimming.

Coordination
This stroke uses constant arm and leg action.

T-Method

The T-method is a basic drownproofing method. This is the best survival technique if you have negative buoyancy. To execute the T-method—

With your face out of the water, take a deep breath and submerge your face in the water while holding your breath.

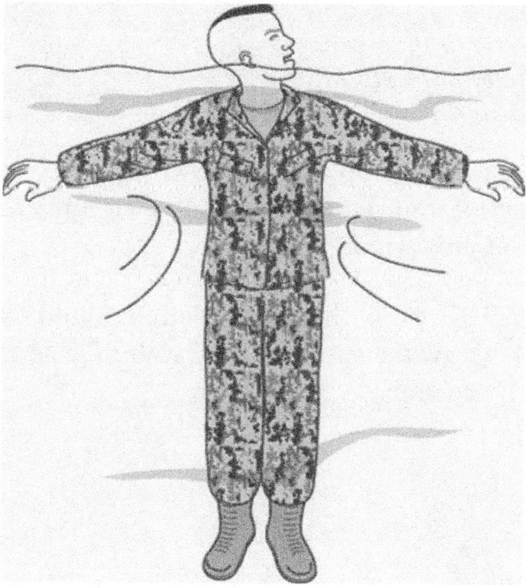

Float with your body in a horizontal position, arms extended from your side, and legs extended and joined.

Move your hands up to your armpits by tracing an imaginary line along your ribs.

Extend your arms outward (horizontal) to your sides, your body position resembles the letter "T."

Step out and forward with one leg and point your other leg to the rear, your knees should be slightly bent. Simultaneously, bring your arms down to your sides. Then bring your legs back together. You complete the steps by exhaling most of your air and preparing to surface your face to obtain another breath of air. Hold your head out of the water. Tilt your head back slightly. Breathe normally.

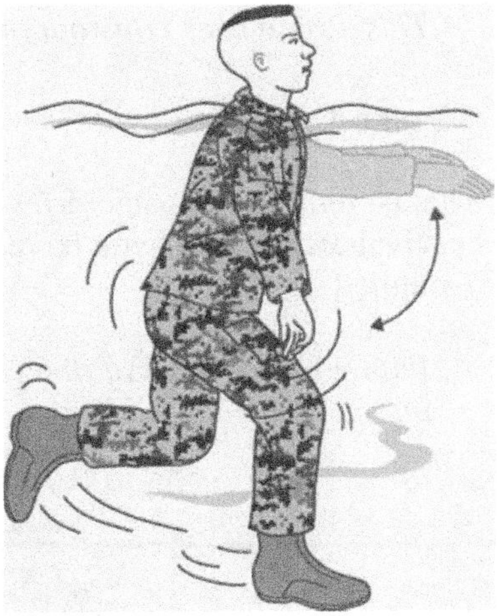

Once your breath is complete, move your hands up and down directly in front of your body. Do this two or three times to slow your descent into the water.

> NOTE: To avoid hyperventilating, hold your breath below the surface of the water for no more than 10 seconds.

The Sweep

The sweep works well if you have slight to excellent positive buoyancy. To execute the sweep—

Float face down in the water: bend 45 degrees at the waist, arms and legs dangling, head hanging down, relax all muscles.

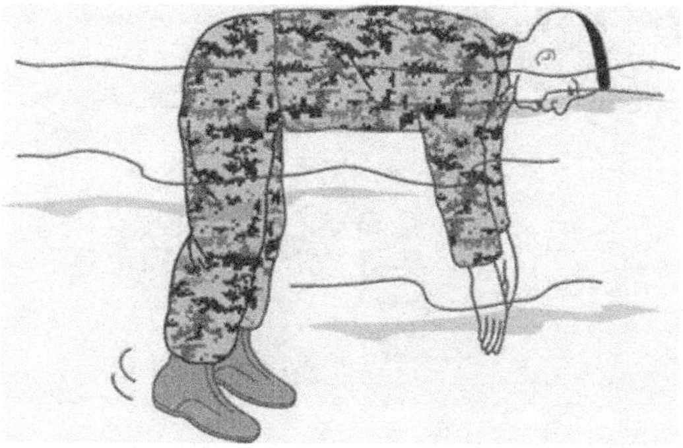

Spread your feet slowly to prepare for a single kick (one leg is forward and one leg is rearward).

Cross your arms in front of your chest, palms outboard with the back of each hand touching the opposite ear.

Exhale prior to raising your head for a breath.

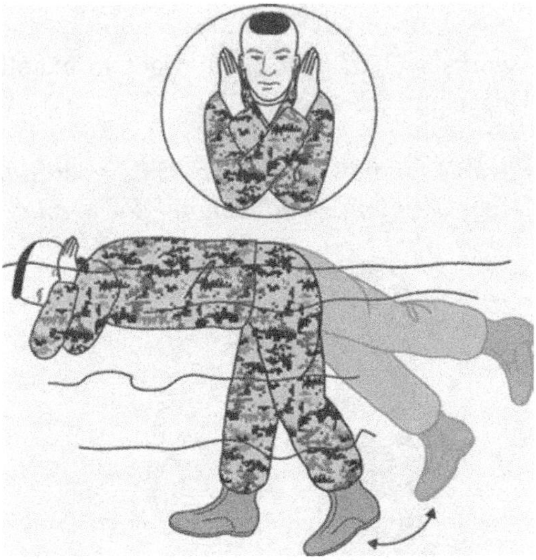

Bring your legs together and sweep your arms down and out until the arms are fully extended out to the sides. This raises your face above the water and allows you to catch a breath of air.

With your air supply replenished, return your face to the water and relax while sweeping your arms in a downward motion in front of your body to prevent/slow your descent. Do not hold your breath for more than 10 seconds. More than 10 seconds enhances your chances of shallow water black out and subsequent drowning.

Breast Stroke

Use this stroke to swim underwater, through oil or debris, and in rough seas. If you are a good swimmer and not wearing combat gear, the breast stroke is the best stroke for long-range swimming because it provides good visibility and allows you to conserve your energy and maintain a reasonable speed.

Body position
Lie prone in the water. Swim with your trunk and legs projecting back and down at an angle of 20 to 30 degrees.

Extend arms out in front (hands together [side by side]), and extend legs behind (toes pointed) to prevent drag.

Face downward, looking forward at a 45 degree angle to break the water and to prevent water from washing into the collar area causing drag.

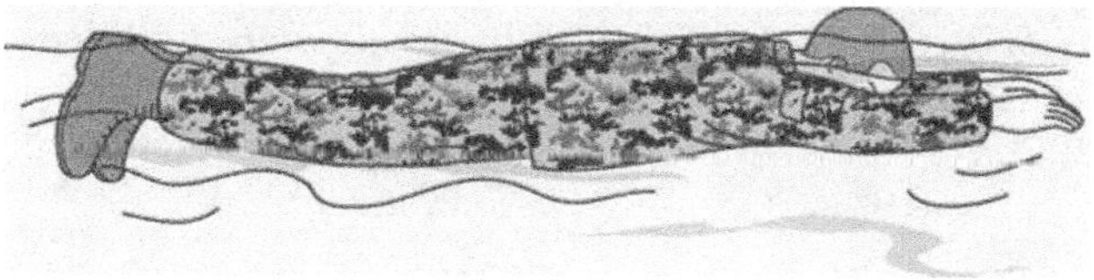

This is known as a glide.

Arm action

Turn your palms outward and bend your arms slightly.

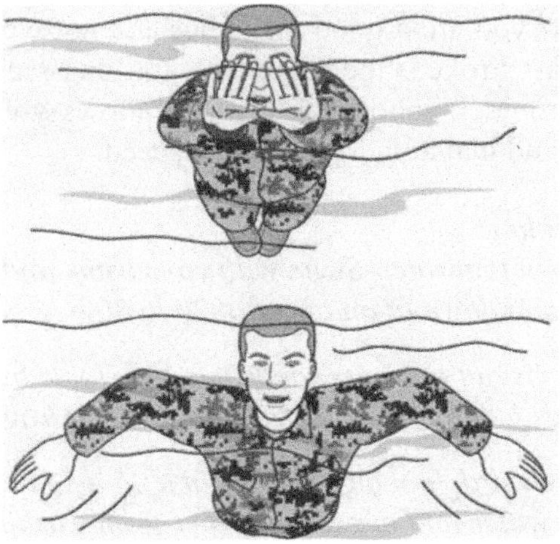

Sweep your arms sideward and slightly downward until your hands are opposite and slightly below your shoulders.

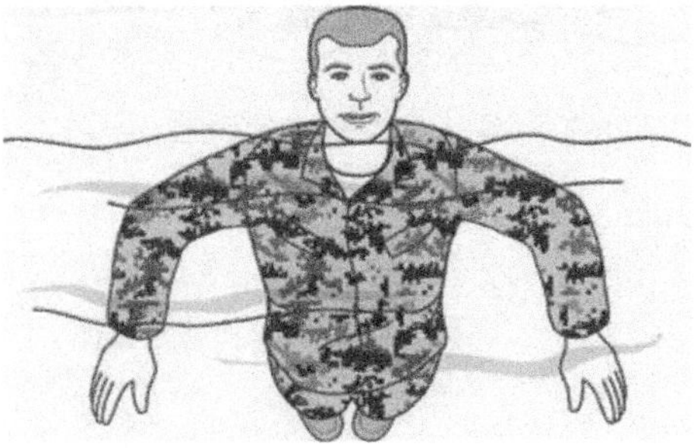

Rotate your head up, breathe once your mouth breaks the surface.

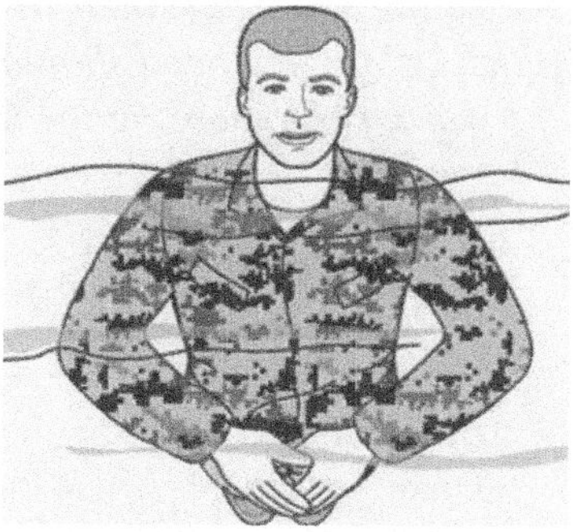

Bring your hands and arms up along your chest and thrust them forward until they are extended and ready to execute the next arm pull.

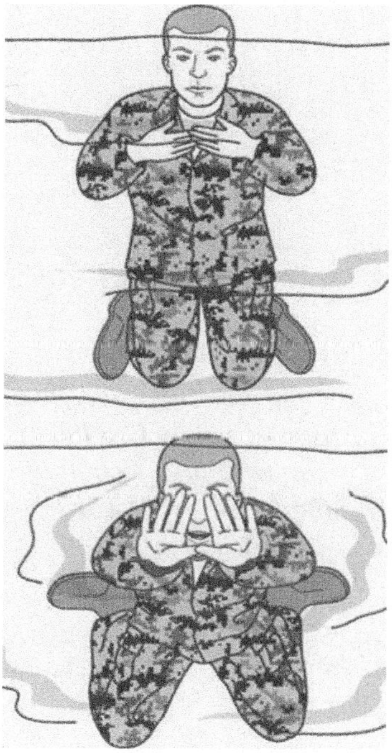

As the arms start their recovery into the glide, the head should rotate forward, resubmerging the face.

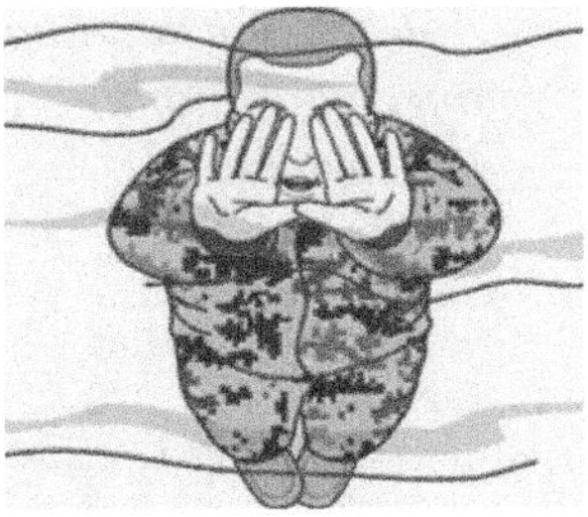

Leg action

Draw your heels toward your buttocks, establish a 45 degree bend in the knees.

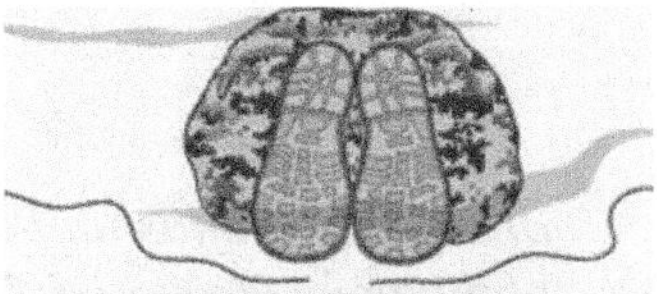

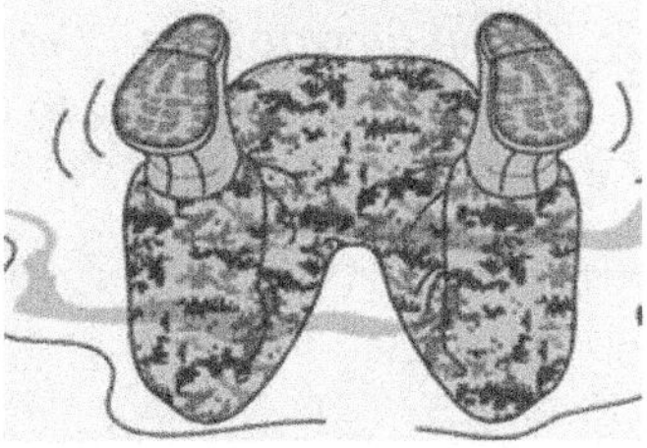

Thrust your legs outward and rearward, then squeeze them together. The whipping action of the feet aids forward propulsion.

This is known as the breast stroke kick.

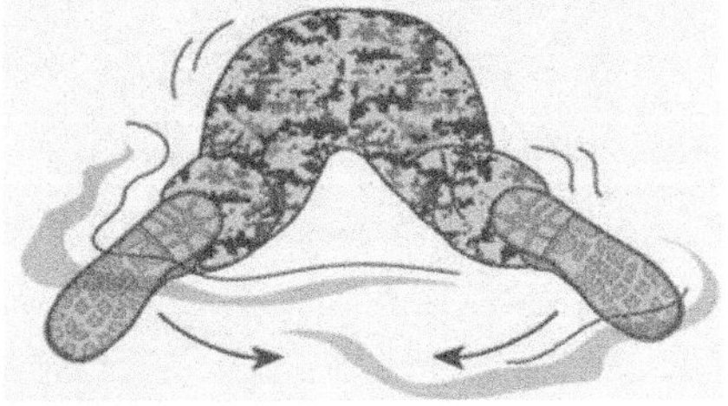

Breathing
Inhale during the arm pull and exhale through your mouth and nose during the finish of the breast stroke kick and glide.

Coordination
The stroke movement is in three counts:

Begin your arm pull. Near the finish of the pull, flex your knees and bring your heels toward your buttocks. The arm pull counteracts the resistance created by the knees.

As the arm pull is completed, thrust your hands forward, kick your legs outward and rearward, and squeeze them together.

Glide through the water for approximately 1 to 3 seconds or until your forward momentum decreases, then begin the next stroke.

Side Stroke

The side stroke is a survival stroke because you use both arms for buoyancy, with each arm creating a slight propulsion. The majority of your body's propulsion comes from your kick. To execute the side stroke—

Body position
Lie on your side with your lead (bottom) arm extended beyond (with a slight bend in your elbow) your head and in line with your body. Palm is down and your hand is submerged 6 to 8 inches.

Extend your trail (top) arm down the length of your body over your thigh.

Keep your legs straight and together, toes pointed rearward.

Keep your face out of the water, this allows for free breathing.

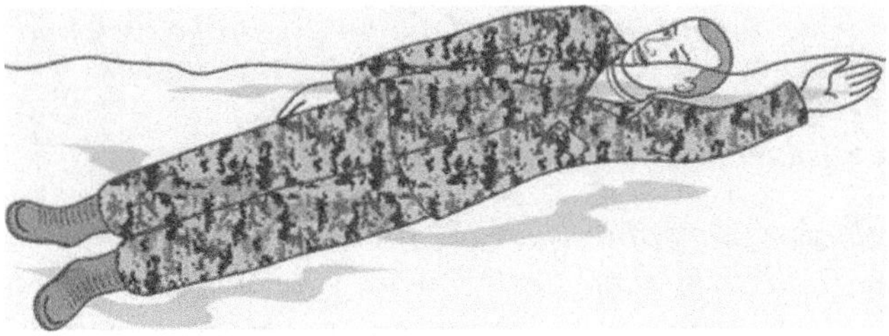

This is known as the glide.

Arm action
With your lead arm, pull your arm downward, while flexing at the elbow, until it is straight down from your shoulder.

Rotate your shoulder and pull your elbow into your side. This should put your lead hand at shoulder level. At the same time, turn your palm toward your face and thrust forward to your original, extended position.

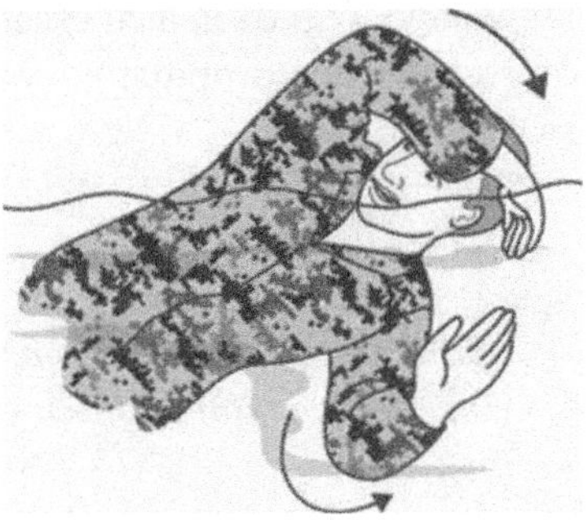

Draw your right hand upward in front of your chest to shoulder level. Rotate your palm toward your feet, then push it downward in front of your body toward your feet to catch the water.

Push your trail hand backward to its original position on top of your thigh. (Your trail hand starts forward and meets your lead hand at your chest/shoulder.)

Leg action

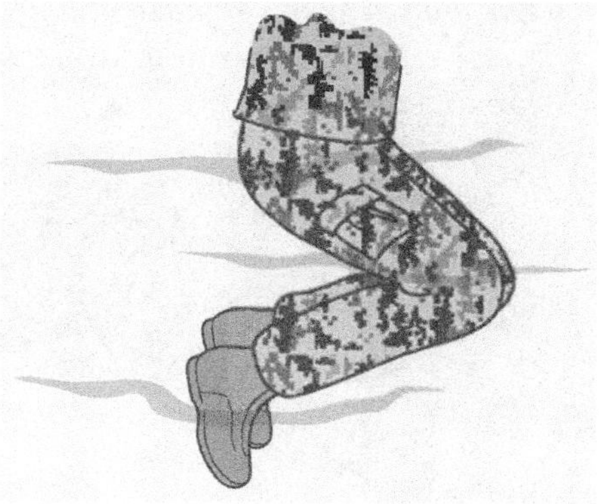

To perform the scissor kick, the top leg always goes forward and the bottom leg always goes rearward. From the extended position, draw—or recover—your feet toward your buttocks until your legs are bent at a 45 degree angle at the knees and the hips are flexed at a 45 degree angle with the thighs.

Once the legs have completed their recovery and while maintaining a 45 degree bend in the knees, extend the legs fully into a "V" shape in order to catch the water for the propulsion phase.

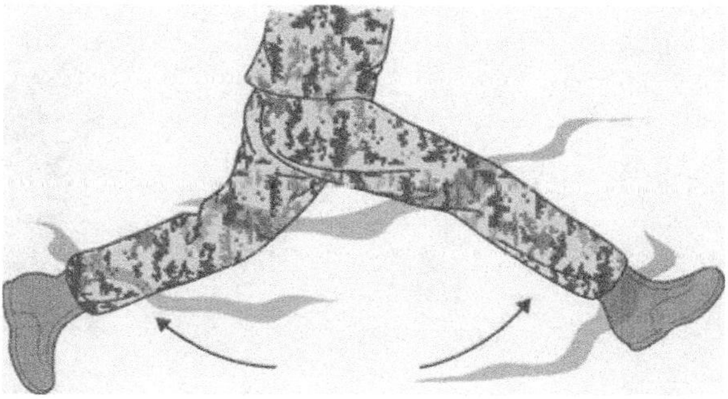

Once the legs are separated and extended forward and rearward to the "V" position, sweep the legs together until the feet are together.

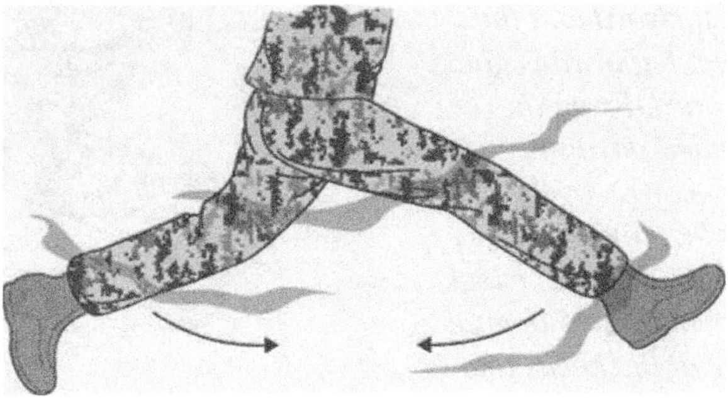

You are now in the glide position.

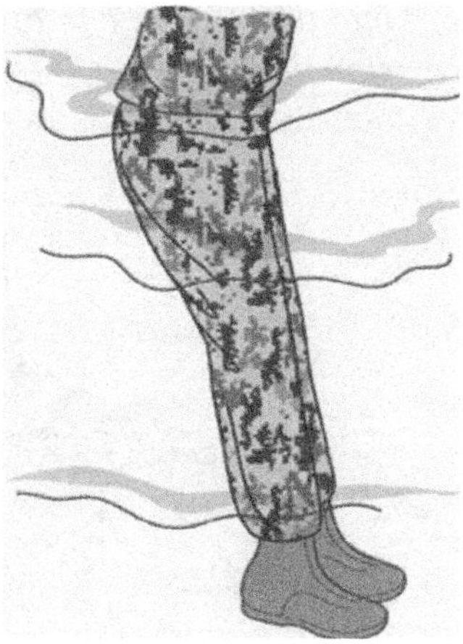

Breathing

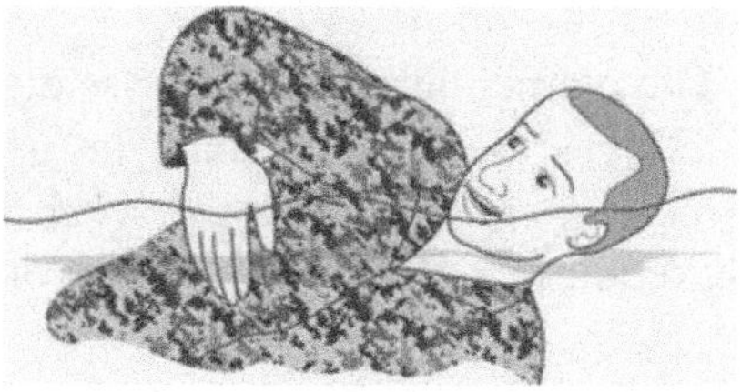

As long as your face remains clear of the water, it is a free breathing stroke. However, it is recommended that you exhale then quickly inhale when the legs are sweeping back together in the scissor kick. This is when the body reaches its highest point in the water, thus clearing the face completely from the surface of the water making it the optimum time to breathe.

Coordination

Begin the stroke with the downward pull of your lead arm. At the same time, bring your trail arm upward and draw your knees up to begin the kick. Let the thrust of the lead arm, push of your trail arm, and the kick of your legs coincide in order to finish the glide position. Glide through the water for approximately 1 to 3 seconds or until your forward momentum decreases, then begin the next stroke.

Elementary Backstroke

The elementary backstroke is also an excellent survival stroke. It relieves the muscles that you use for other strokes, and it is the recommended stroke for weak swimmers or nonswimmers. To execute the elementary backstroke—

Body position
Start on your back.

Face up, chest up, and hips up, keeping an arch in your lower back with arms pressed to your sides and your legs extended and joined to prevent drag.

Arm action
Trace your hands up your sides to an area near your armpits then extend your arms out to the sides to form the letter "T" (palms facing feet), locking out the elbows.

> NOTE: Don't raise your arms above your head. This creates drag, changes your body position, and submerges the head.

Slap your palms to your thighs using a strong sweeping motion.

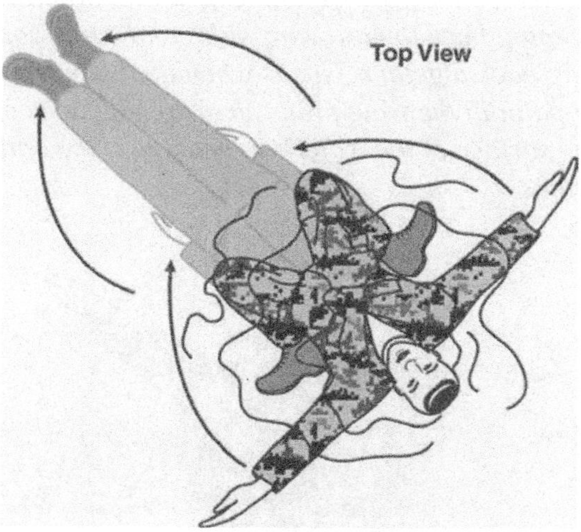

Leg action

Bend both legs at the knee (90 degree angle) slightly separating your knees and drawing your heels downward to a point under and outside your knees. The knees are spread as wide as the hips or slightly wider depending on the body type of the swimmer.

Circle around in a whipping action, ending with legs in a glide position.

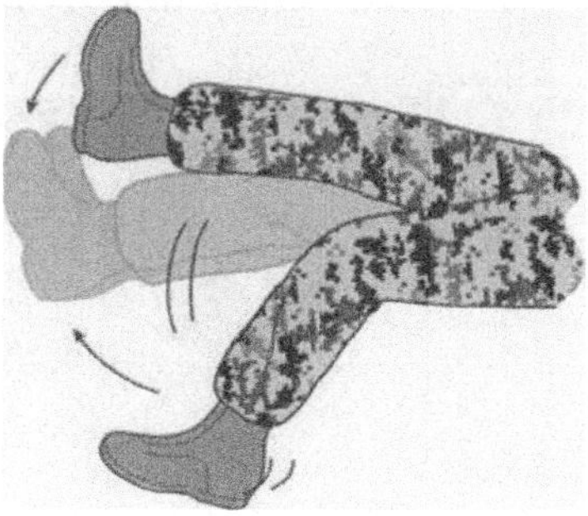

Breathing

Breathe anytime during this stroke. However, it is recommended that you exhale then quickly inhale when your arms are sweeping back toward your sides and while your legs are sweeping back together. This is when the body reaches its highest point in the water, thus clearing the face completely from the surface of the water and making it the optimum time to breathe.

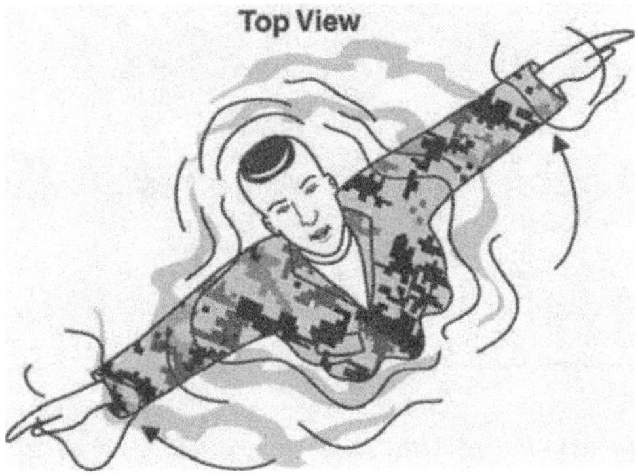

Coordination

The stroke movement occurs in three counts (recovery, catch, power).

Begin the arm pull (recovery).

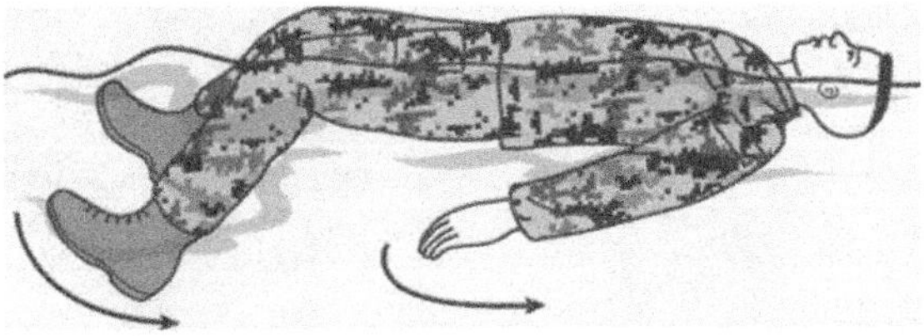

Near the finish of the pull, flex your knees to a 90 degree angle. The arm pull counteracts the resistance created by the knees.

Kick out your legs, and squeeze them together as the arm pull is completed (catch, power).

Glide through the water for 1 to 3 seconds or until your forward momentum decreases, then begin the next stroke as your momentum slows.

Chapter 2
Water Rescues

A drowning victim can panic and react with unexpected violence and can seize and inadvertently drown a rescuer. Therefore, if possible, a water rescue should be executed from a distance. Reaching, wading, or throwing methods are used in a distance rescue.

If a victim is too far away to use these methods, a swimming rescue may be necessary. Remove all combat gear, when possible, before entering the water. Swim within 2 to 6 yards of the victim to maintain a margin of safety, this allows you to reassess the situation and reassure the victim. If the victim is unconscious, use the wrist tow method or cross-chest carry method to pull the person to safety. If the victim is struggling, use a rear approach and then execute either a single armpit level off or a double armpit level off before towing the victim to safety.

If the victim does begin to overpower you, there are techniques that allow you to defend yourself without having to abandon the rescue. These techniques include the block, the wrist-grip escape, the front head-hold escape, and the rear head-hold escape. These techniques allow you to separate yourself from the victim, reassess the situation, and then attempt the rescue again.

> *Note:* In this chapter's illustrations, the rescuer is shown without a helmet.

Reaching Rescue Techniques

Reach

From a safe position at the water's edge, reach out to the victim. Talk constantly to calm the victim. Retain partial contact with land or some solid support structure (e.g., pier, bridge). If the victim is close but still beyond reach, extend an object (e.g., stick; pack; rifle with magazine removed, chamber empty, and muzzle pointing toward victim) that the victim can grasp. Pull the victim slowly to safety. Once the victim is close to shore, remove the victim from the water.

Note: You can also extend a foot to the victim if you can retain a secure grip on a solid support structure.

Reach From a Deck

Reach from a deck rescue can be executed by either a swimmer or a nonswimmer, and it can be used on an active or a passive victim. Be sure to reassure the victim during the rescue. To execute reach from a deck—

Lie prone on the deck with your body firmly anchored. To anchor your body, lie flat, spread legs apart, and extend one arm behind you with your palm down and on the deck.

Keep as much of your weight on the deck as possible and extend your free hand to the victim.

Grasp the victim's wrist from above, your thumb and index finger are facing you. The victim should never be allowed to grab you and put your life at risk. Therefore, never reach across the victim to grab his wrist, always grab the wrist that is the closest to you.

Keep your arm straight and locked out and pull the victim to the side of the deck. The victim should never be allowed to grab you and put your life at risk. Therefore, never pull the victim into you, always pull the victim into the side.

Arm Extension

An arm extension rescue can be used if you cannot reach the victim using the reach from a deck rescue technique and you must enter the water. An arm extension rescue technique is used for a victim who is either active or passive. Once you determine that a reach from a deck rescue technique is not viable—

Reassure the victim and quickly ease into the water while holding onto the deck with one hand.

Grasp the victim's wrist from above, your thumb and index finger are facing you. The victim should never be allowed to grab you and put your life at risk. Therefore, never reach across the victim to grab his wrist, always grab the wrist that is the closest to you.

Keep your arm straight and locked out and pull the victim to your side. The victim should never be allowed to grab you and put your life at risk. Therefore, never pull the victim into you, always pull the victim into the side.

Leg Extension

If the victim is beyond the reach of your arm, ensure that you have a firm grip on the deck, extend a leg to the victim and allow him to grab it, slowly bring the victim in closer until you can grab the victim's wrist that is holding onto your leg. Once you have a firm grasp on the victim's wrist, use the steps in the arm extension to pull the victim to safety. The leg extension rescue can only be used to rescue an active victim because the victim must be able to grab the rescuer's leg.

Wading Assist

Do not wade into water that is deeper than your chest. Talk constantly to calm the victim. If possible, do not touch the victim directly. Extend an object (e.g., stick; pack; rifle with magazine removed, chamber empty, and muzzle pointing toward victim) that the victim can grasp. Once the victim grasps the object, pull the victim slowly to safety.

Throw

If the victim is not within reach, use an expedient line to throw a lifesaving device to the victim. A lifesaving device can be any weighted item that floats (e.g., a canteen that is one fourth full of water). The lifesaving device must be secured to the end of the rope so that the rope will feed out from its coil when tossed. Talk constantly to calm the victim. Once the victim grasps the line or the lifesaving device, pull the victim toward you at a steady pace that keeps the victim's head above the water's surface. DO NOT pull so strongly that you break the victim's grip on the line. To prepare and use an expedient line and lifesaving device—

Tie a bowline at one end of the rope. (The appendix illustrates various types of knots.)

Unfasten the lid of a canteen.

Place the bowline around the neck of the canteen.

Refasten the lid so the canteen hangs from the bowline loop.

Place one end of the rope under the ball of your forward foot to secure it, either tie a knot at the end of the rope or tie an object at the end of the rope to create a block. Stand with your weight on the end of the rope. DO NOT tie the rope around your ankle.

Coil 20 to 30 yards of the rope, and hold it in a nonthrowing hand.

Place the canteen in your throwing hand.

Use an underhand throw to pitch the canteen and rope a short distance over the victim's head. Keep your nonthrowing hand open so the coil can unfold freely. The rope should trail across the victim's outstretched arms.

Retrieve the rope if the throw is inaccurate or the victim fails to grasp it. Recoil the rope as it is retrieved.

Divide the coil and throw again.

Lifesaving Approaches

Properly approaching the victim is as important as any other aspect of a rescue. Knowing the appropriate and safest method needed to approach a victim is critical in ensuring your safety and the victim's survival. Determining the victim's physical state (distressed swimmer, victim that is active and drowning, or victim that is passive and drowning) is crucial and will determine which type of approach you will execute.

Front Surface Approach

The front surface approach is typically performed when the victim is passive. You must remember that approaching the victim from the front is dangerous because a distressed swimmer or victim that is active can lunge toward you.

Rear Approach

If the victim is active, you must remember that his response can change rapidly. Therefore, approaching him from the front could be extremely dangerous. The victim could easily grab you and take you under the water during a state of panic. Therefore, use a rear approach when possible. When approaching a victim from the rear, reassure the victim until you get to within 2 to 6 yards of the victim; at that point, stop talking to the victim in order to maintain an element of surprise.

Approach Strokes

Before approaching a victim, you must evaluate the victim:

- Is the victim a distressed swimmer?
- Is the victim active and drowning?
- Is the victim passive and drowning?

Select your approach stroke based on your evaluation. Remember, the safety of the rescuer is paramount. Do not endanger yourself in an attempt to reach a drowning victim.

Crawl Stroke Approach Stroke

The crawl stroke (free style) is the fastest approach stroke. This stroke is used when the victim is passive and/or unconscious in the water. That is, the victim is face down, submerged or near the surface of the water, is not breathing, or is not moving. If any of these conditions exist, there is an obvious need to reach the victim as quickly as possible.

Execution of the crawl stroke approach stroke is the same as the crawl stroke (see pages 1-40 through 1-44), except that your head remains above the water's surface to allow for free breathing and maintaining eye contact with the victim.

Maintaining eye contact is critical. If the victim becomes submerged, you will have a better chance of locating him if you know his last location. Your distance from the victim will determine if your head remains out of the water the entire time you are executing the stroke. If you are 54 yards or less from the victim, you should maintain eye contact by keeping your head raised until you reach the victim. If you must swim more than 54 yards to reach the victim, make eye contact with the victim, place your face back in the water for four or five strokes, then raise your head again to regain eye contact with the victim (repeat these steps until you reach the victim).

Body position
The position of your body is horizontal to diagonal and influenced by the position of the head. This allows you to maintain eye contact with the victim.

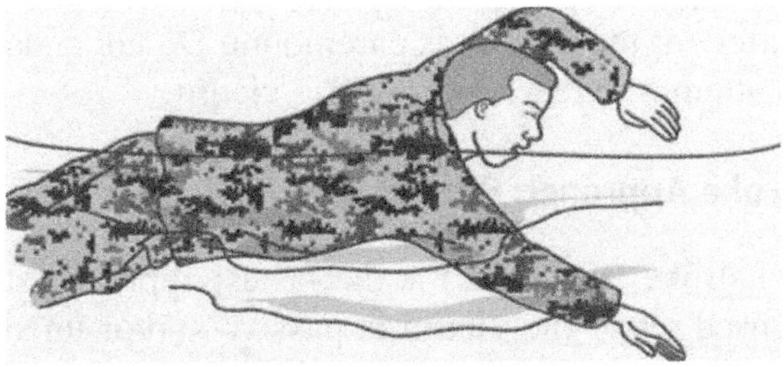

Arm action
While one arm is forward of your head to catch the water for the propulsion phase, extend your other arm alongside your body in what is known as the recovery phase. The recovery phase ends when your arm is extended forward of your head to begin the propulsion phase. In the propulsion phase, your arms are approximately shoulder-width apart and the pulling action is slightly wider and deeper to compensate for your raised head.

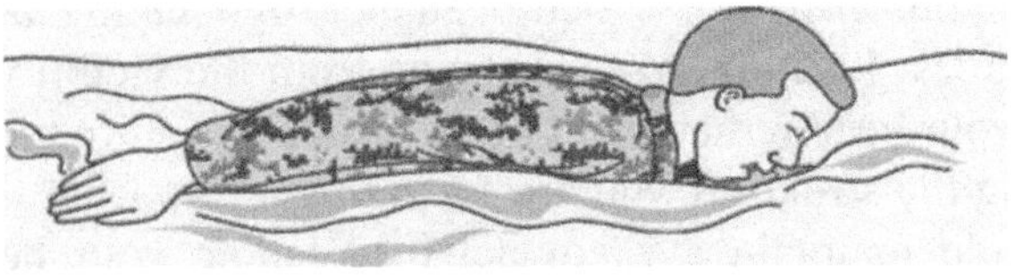

Leg action
Kick one leg in a downward motion while your other leg is recovering to the surface to prepare for the next kick (this is known as a flutter kick). This kick is used for propulsion and maintaining the lower body as horizontal to the water's surface as possible. When your head is raised, your knees are slightly bent in order to keep your feet near the surface.

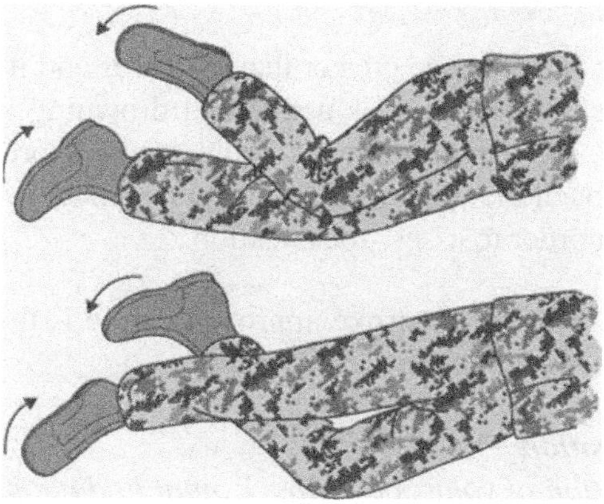

Breathing
If your face is above the water, breathe as needed (free breathing).

If traveling in excess of 54 yards to reach the victim, rotate your head to the right or left during the recovery phase and inhale when your face is out of the water or wait until your head is out of the water and facing forward while regaining sight of the victim.

Coordination
This stroke uses constant arm and leg action.

Breast Stroke Approach Stroke

The breast stroke approach stroke is used when a swimmer is distressed or a victim is active and drowning. This stroke is also used when the victim has a suspected spinal injury because it minimizes the wake created around the victim, minimizes the movement of the victim's head and neck, and helps prevent further injury.

If a victim is at a distance greater than 54 yards and if the swimmer is in distress or the victim is active and drowning in open water (ocean, lake), you should use the crawl stroke approach stroke to rapidly approach the victim, stopping approximately 11 yards from the victim in order to assess the situation.

Execution of the breast stroke approach stroke is the same as the breast stroke with a few minor deviations.

Body position
The position of your body is horizontal to diagonal and influenced by the position of the head. This allows you to maintain eye contact with the victim. It also allows you to communicate with the victim in order reassure him as you approach.

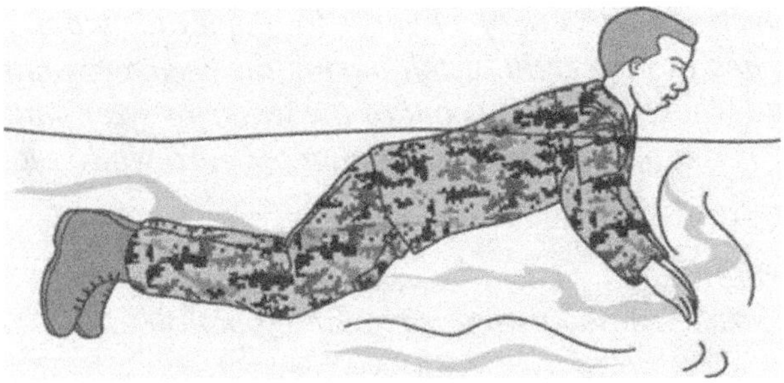

Arm action

Execute the arm action for the breast stroke, except that your arm pull is wider and deeper, which allows your head to remain above the water's surface.

Leg action
Execute the leg action for the breast stroke, except that your legs open wider during the propulsion phase of the stroke.

Breathing
Keep your face above the water and breathe as needed (free breathing).

Coordination
Coordination is the same as the breast stroke, except that there is no glide period. You constantly stroke until you reach the victim.

Level Offs

A properly performed level off positions the victim's face up and horizontal to the water's surface. There are three types of level offs: front surface approach, single armpit level off, and the double armpit level off. The single armpit level off and the double armpit level off are done by performing a rear approach, which is the preferred approach when a victim is active.

Front Surface Approach

The front surface approach is performed when the victim is passive. Approaching a victim from the front may place you in danger because a distressed swimmer or a victim who is active may lunge toward you. To execute a front surface approach—

Stop 2 to 4 yards from the victim to reassess the situation and reassure the victim.

Determine which arm to use in order to rotate the victim into a face-up position (your right hand on the victim's right wrist or your left hand on the victim's left wrist).

Move into position and prepare to grasp the victim.

Turn sideways and move toward the front of victim.

Once in position, reach forward and grab the victim's wrist. Your thumb is on the inside of the victim's wrist, as if you were checking his pulse with your thumb. Your remaining fingers wrap around the victim's wrist.

Lean back immediately and execute a powerful scissor kick or inverted scissor kick and perform short, vigorous pulls with your free arm.

As the victim begins to move forward in the water, kick, pull, and twist outboard on the victim's wrist. The momentum created from the kick and the pulling and twisting action of the victim's wrist will rotate the victim into a face-up and horizontal position in the water.

Extend and lock out your towing arm down the length of your body and execute a wrist tow to move the victim to safety.

Single Armpit Level Off

Approach the victim.

Stop 2 to 6 yards from the victim to reassess the situation, reassure the victim, and maintain your margin of safety.

Approach the victim slowly and grab his armpit with your hand (your right hand to victim's right armpit or your left hand to victim's left armpit). Position yourself sideways to the victim and place your elbow in the center of his back.

Pull with your hand that is in the victim's armpit while pushing with your elbow that is in the victim's back to place the victim horizontal on the water's surface (face up). To assist in placing the victim horizontally, use your free arm to execute short, vigorous pulls and use your legs to execute a scissor kick or inverted scissor kick.

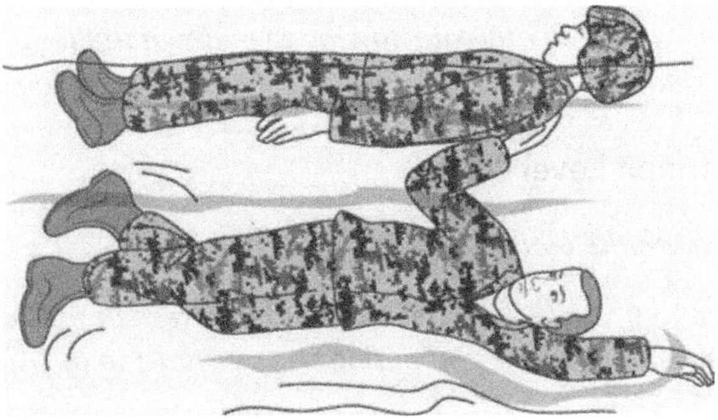

Once the victim is horizontal and on the water's surface, begin forward momentum by extending your towing arm to a fully locked-out position and executing a single armpit tow.

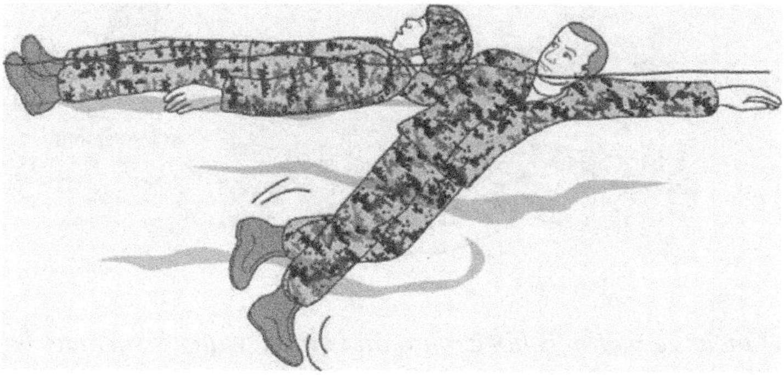

Double Armpit Level Off

Approach the victim.

Stop 2 to 6 yards from the victim to reassess the situation, reassure the victim, and maintain your margin of safety.

Approach the victim slowly and grab his armpits with both your hands. Place your elbows on his back.

Pull with your hands that are in the victim's armpits while pushing with your elbows that are in the victim's back and use an inverted breast stroke kick to place the victim horizontal on the water's surface (face up).

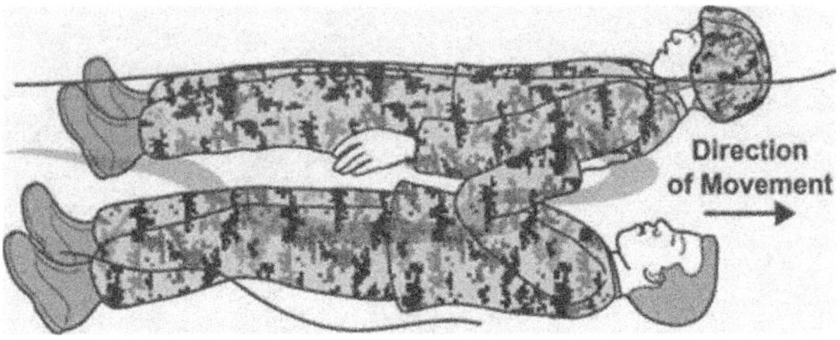

Once the victim is horizontal and on the water's surface, begin rearward momentum by slowly extending your arms into a fully locked-out position and executing a double armpit tow and inverted breast stroke kick.

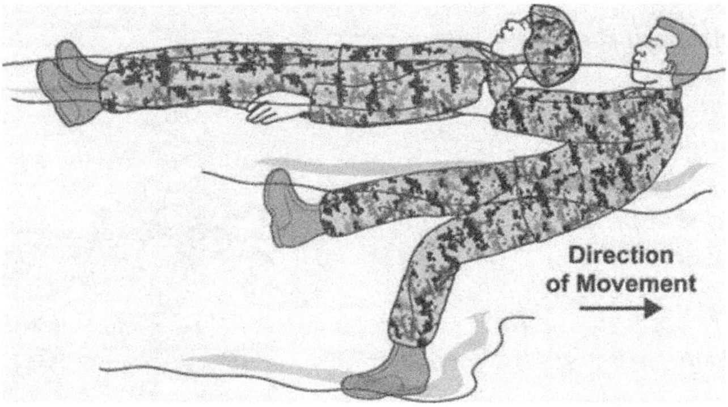

Rescue Techniques

Wrist Tow

Use the wrist tow method to rescue a victim who is floating face down. DO NOT use the wrist tow on a struggling victim. If time allows, remove your helmet and gear before attempting the rescue. Swim toward the victim using a modified breast stroke. Swim within 2 to 6 yards of the victim to maintain a margin of safety, this allows you to reassess the situation and reassure the victim. The following steps show proper front surface approach, wrist tow procedures:

Approach the victim from the front and grasp the underside of the victim's left wrist with your left hand or the right wrist with your right hand. Ensure that your thumb is on the underside of victim's wrist.

Lean back, pulling and kicking strongly to move the victim into a horizontal position.

Twist the victim's wrist to rotate the victim into a face-up position.

Swim toward safety using the lifesaving stroke.

> NOTE: The lifesaving stroke is a modified side stroke wherein the top arm is used to tow or carry a victim to safety and use a scissor kick or an inverted scissor kick is used for propulsion.

Keep a firm grip on the victim's wrist.

Keep your towing arm fully extended and along your side. This keeps the victim in the water column and prevents drag.

Ensure the victim's head does not go under water during the recovery.

Single Armpit Tow

You perform the single armpit tow after a victim has been properly leveled off. The single armpit tow uses the lifesaving stroke. To execute the single armpit tow—

Place one hand (top arm) in the victim's armpit. Your towing arm is straight and along your side.

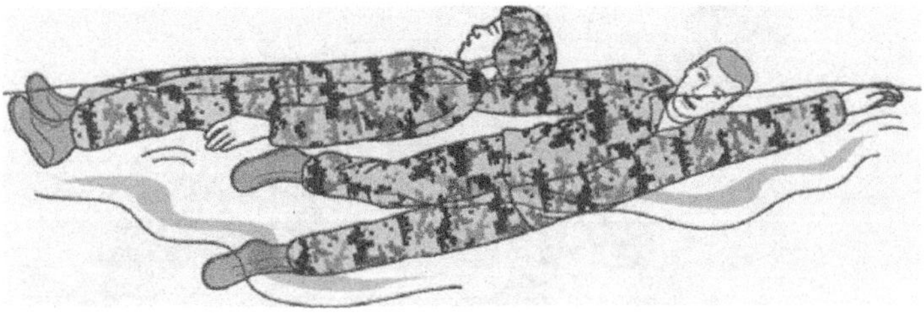

Use your lead arm (bottom arm) to execute short, vigorous pulls.

Execute a scissor or inverted scissor kick in a continuous and vigorous manner.

Use either free breathing or explosive breathing during the tow. During free breathing, keep your head above the water's surface and continuously reassure the victim. During

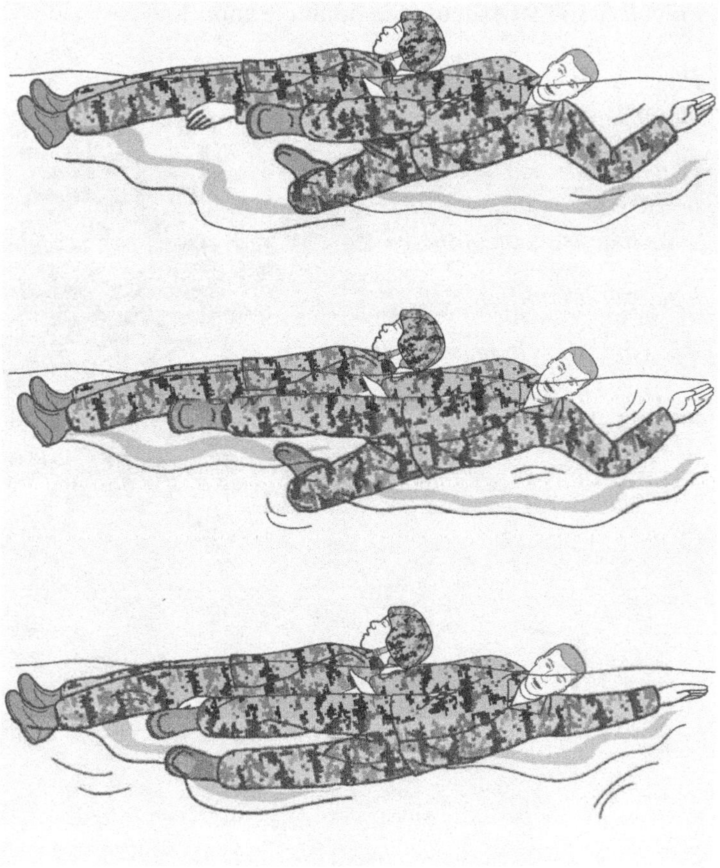

explosive breathing, put your head under the water to plane off your body's angle and reduce drag. Execute approximately two strokes, then lift your head up to breath and reassure the victim. Return your head back underwater for the next few strokes.

Double Armpit Tow

You perform the double armpit tow after a victim has been properly leveled off. To execute the double armpit tow—

Place both hands in the victim's armpits.

Extend both arms fully along your body. You are on your back, and your face is clear of the water.

Use free breathing.

Reassure the victim at all times.

Use the inverted breast stroke kick (it is the only kick that can be used due to your body position) to tow the victim to safety. (The inverted breast stroke kick is the same kick used in the elementary backstroke.) Your kick must be continuous and vigorous in order to keep the victim's face above the water.

Collar Tow

You perform the collar tow after a victim has been properly leveled off. To perform a collar tow using the victim's blouse—

Maintain control of the victim by grasping his armpit with one hand and then with your free hand grasp either his combat gear or his blouse between his shoulder blades. If grasping the blouse, grasp the material with your palm up, then turn your hand over to tighten the material.

Release the victim's armpit once control is established and execute the lifesaving stroke to tow the him to safety.

Cross-Chest Carry

Use the cross-chest carry to carry a victim to safety if the victim is struggling or when moving through heavy surf. Remove your helmet and gear before attempting a rescue. Talk constantly to calm the victim. Swim toward the victim using a breast stroke approach stroke. Swim within 2 to 6 yards of the victim to maintain a margin of safety, this allows you to reassess the situation and reassure the victim. The following steps illustrate proper cross-chest carry procedures:

> **CAUTION:** The cross-chest carry causes fatigue even if you are in excellent physical condition.

Use a level-off technique to place the victim in a horizontal, face-up position.

Retain a grip on the victim with one hand. Reach over the victim with your free hand to encircle the victim's chest. Place your free hand on the victim's opposite rib cage, just below his armpit.

Release your grip once you have a secure hold on the victim's chest.

Swim toward safety using the lifesaving stroke while keeping a firm grip on the victim's chest and your hip on his back.

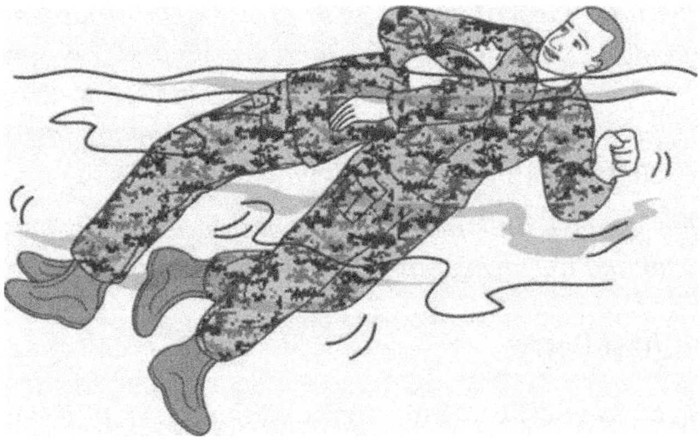

This procedure brings the victim's face and shoulders clear of the water, and typically the victim stops struggling. Sometimes, however, the victim will struggle during the swim to safety. If this happens, either tighten your grip on the victim or defend yourself with one of the techniques discussed on pages 2-25 through 2-30.

Tired Swimmer's Assist

You must maintain a 2 to 6 yards margin of safety from the victim at all times while you are getting into position to perform the tired swimmer's assist. Once behind the victim, you extend one arm, hand straight. Place your straight hand underneath the victim's opposite armpit. Maintain a 45 degree angle away from the victim, your arm is locked out. Assist the victim in propulsion until both of you reach safety.

Defense Against a Drowning Victim

DO NOT sacrifice your life in an attempt to save the victim. A struggling drowning victim poses great danger to anyone nearby. Driven by panic, the victim can grab you with great strength in an effort to climb out of the water. This can result in death for both you and the victim. The following techniques can be used to defend against a drowning victim, but the best defense against attack by the victim is to stay out of reach.

Block

The block prevents the victim from grabbing you if you have approached from the front. If the victim lunges toward you, react as follows:

Place one or both open hands against the victim's upper chest; being careful to avoid the victim's face, neck, and abdomen.

Lean backwards and submerge rapidly. Keep your blocking arm(s) extended.

Swim underwater and away from the victim, and return quickly to the surface.

Stop 2 to 6 yards from the victim to reassess the situation.

Determine an appropriate course of action.

Wrist-Grip Escape/Wrist-Grip Escape Alternative

The wrist-grip escape is used when a victim grabs your arm or wrist. Submerge the victim quickly by reaching across with your free hand, pushing down on the victim's shoulder to submerge him, and kicking to propel yourself upward. While keeping the victim submerged with your hand on his shoulder, give three hard jerking pulls with your trapped hand in an attempt to break free from his grasp. Once free, swim clear of the victim and reassess the victim's condition.

If the wrist-grip escape is unsuccessful in freeing your hand, use the wrist-grip escape alternative: use your free hand to grab your trapped fist, rotate thumbs up, apply bone-on-bone contact with the victim's arm, pry your hand out of his grasp, and quickly swim away from the victim.

Front Head-Hold Escape

The front head-hold escape allows you to escape when you are facing a victim who is gripping you around your head and neck. To execute the front-head hold escape—

Take a quick breath and tuck your chin into your shoulder to protect your throat.

Clap your hands above your head (three times) to submerge instantly. This drags the victim below the water, lifts his arms from around your neck, and, typically, he releases his grasp in order to get back to the surface. If he doesn't release his grasp, apply pressure to the victim's brachial pressure points (which are located inside of the upper arm, above the elbow).

Thrust the victim's arms up and away.

Keep your chin tucked to protect your throat, and swim underwater away from the victim and return quickly to the surface at the ready position.

Stop 2 to 6 yards from the victim to reassess the situation.

Determine an appropriate course of action.

Rear Head-Hold Escape

The rear head-hold escape allows you to escape when a victim is gripping your head and neck from the rear. To execute the rear head-hold escape—

Take a quick breath. and tuck your chin down, turn your head to either side, and raise your shoulders to protect your throat.

Take a strong stroke, clap your hands above your head (three times), and submerge instantly. This drags the victim below the water and, typically, he releases his grasp in order to get back to the surface. If he doesn't release his grasp, apply pressure to the victim's brachial pressure points (which are located inside of the upper arm, above the elbow).

Thrust the victim's arms up and away.

Twist your head and shoulders until free.

Swim underwater away from the victim and return quickly to the surface.

Stop 2 to 6 yards from the victim to assess the situation.

Determine an appropriate course of action.

Administering First Aid/Rescue Breathing

To administer first aid/rescue breathing, remove the unconscious victim from the water when possible and stop any bleeding.

First, open the airway and check for breathing. If the victim is not breathing, give two breaths of air. Check for a pulse. If there is a pulse but the victim is not breathing, continue rescue breathing. If there is no pulse, start cardiopulmonary resuscitation (CPR). For detailed information on administering CPR and rescue breathing, refer to American Red Cross CPR skill cards.

WARNING

If the victim has no pulse and is not breathing, administer CPR immediately. If the victim does have a pulse but is not breathing, give rescue breathing only. If the victim has a pulse and is breathing, DO NOT give CPR—CPR could prove fatal.

Next, protect any wounds from exposure, and treat for shock.

Chapter 3
Treatment of Casualties and Avoidance of Dangerous Marine Life

To survive in the water, you will face many challenges other than just being able to swim. For example, injuries and/or fatigue can lead to drowning and exposure to the elements can lead to hypothermia or heat injuries and may require medical attention. Finally, your presence in the water may be seen as a threat to or food source for a variety of marine life and you may be attacked. This chapter advises you on how to minimize these risks in order to stay alive.

Drowning

Drowning is suffocation by a liquid. A drowning victim inhales water into the lungs or the throat closes by reflex so that little or no water can enter the windpipe. In either case, a victim can no longer breathe.

Symptoms

One symptom of drowning is that the victim may call for help and has an expression of dread or panic. But typically a victim that is active and drowning may not call for help because he is trying to conserve his air and will not speak.

Another symptom of drowning is when the victim thrashes at the water's surface. If the victim stops or grows calmer, he has likely been overcome by fatigue, hypothermia, or a lack of air. At this stage, the victim usually has 20 to 60 seconds before going under the water's surface.

Treatment

If the victim is not breathing, begin rescue breathing. Place the victim on his back, tilt head back to open airway, pinch the nose, and give two full breaths. If the victim does not inhale during the first two breaths, reposition his head and attempt two more breaths. Check for a pulse. If a pulse is present, but the victim is still not breathing, continue rescue breathing. If a pulse is not present, begin CPR. See MCRP 3-02G, *First Aid*, for rescue breathing and CPR details.

WARNING

If the victim has no pulse and is not breathing, administer CPR immediately. If the victim does have a pulse but is not breathing, give rescue breathing only. If the victim has a pulse and is breathing, DO NOT give CPR—CPR could prove fatal. Continue first aid until medical help arrives.

A victim who is not breathing and has no pulse may appear dead. DO NOT decide that death has occurred. Continue with the prescribed treatment. A corpsman or medical officer should decide whether the victim can or cannot be revived.

Hypothermia

Hypothermia is the abnormal lowering of the body's internal (or core) temperature to 95 degrees or below. It occurs when the body loses heat faster than it produces heat. A hypothermia victim loses the ability to move quickly, becomes mentally sluggish, slips into semiconsciousness, lapses into a coma, and dies when internal body temperatures drop too low.

The chilling effects of cold air, wind, or water can produce hypothermia. Water poses the greatest threat because it transfers heat 25 times faster than air. Depending on the water's temperature, a victim can succumb to hypothermia within a few minutes. The body's sudden contact with cold water can also set off a body reaction known as the mammalian diving reflex. This reflex can greatly increase survival time (especially for women and children) in or under cold water. The mammalian diving reflex shuts off blood circulation, except for the flow between the heart, lungs, and brain. The small amount of oxygen left in the blood and lungs is saved for the body's vital organs. This reflex has allowed people to survive being under cold water for an extended period of time. Therefore, a cold water drowning victim should be treated as if still alive even though the victim is not breathing, has no pulse, and may appear dead. DO NOT decide that death has occurred. Continue with the prescribed treatment. Victims of hypothermia can appear to be dead when they are not. A corpsman or medical officer should decide whether the victim can or cannot be revived.

WARNING

If the victim has no pulse and is not breathing, administer CPR immediately. If the victim does have a pulse but is not breathing, give rescue breathing only. If the victim has a pulse and is breathing, DO NOT give CPR—CPR could prove fatal. Continue first aid until medical help arrives. Check for a pulse for at least 45 seconds.

Symptoms

Once the body's core temperature drops, the victim will show one or more of the following symptoms:

- Violent and uncontrollable shivering as the body tries to warm itself.
- Slow or slurred speech.
- Disorientation or poor coordination.
- Loss of skin color.
- Blue and pinched lips.
- A slowing or stopping of shivering that progresses into a rigid torso and limbs.

Survival Time

A hypothermia victim's survival depends on the water's temperature and the time spent in the water. A small body build cools faster than a large build. Children cool faster than adults.

To increase your chance of survival in the water, utilize the HELP position described on page 1-37. Extra clothing and inactivity (remaining motionless in the water) can also increase your survival time.

Treatment

A hypothermia victim must be warmed to prevent further heat loss; therefore, treatment should begin as soon as possible. Consciousness of the victim determines the treatment that should be pursued. See MCRP 3-02G for specific treatment information. The following treatment procedures are recommended:

- If the victim is conscious, give the victim warm fluids. Give candy or sweetened foods to a victim who is able to eat.
- If the victim is unconscious, place him on his back with his head tilted back to ensure an open airway.
- DO NOT massage the victim. Massage can break blood vessels and create swelling, internal pressure, and blocked blood circulation.
- DO NOT give alcohol to the victim. Alcohol lowers the victim's body temperature.
- Shock is a possibility, treat accordingly.
- Seek medical help immediately.

If you are able to remove the victim from the water, apply the following steps when possible:

- Get the victim into shelter.
- Remove the victim's wet clothing.
- Put the victim in dry clothing.
- Place the victim in a sleeping bag if one is available. It may be necessary to place another Marine in the sleeping bag with the victim.
- Place as much insulation as possible between the victim and the ground.
- Use hot water bottles, electric blankets, or blankets heated in an oven or by a campfire to warm the victim's neck, groin, and the sides of the chest.

CAUTION: DO NOT apply heat to extremities.

Heat-Related Injuries

Heat-related injuries include heat cramps, heat exhaustion, and/or heat stroke. They can occur when a Marine—

- Is exposed to extreme heat, such as from the sun or a combination of high air temperatures and water temperatures.
- Does not wear proper clothing or gets overexposed to the sun.
- Becomes dehydrated.

With heat cramps, muscles may cramp in the arms, legs, and/or stomach. The victim may sweat excessively. To treat, create improvised shade for the victim and have him drink water.

With heat exhaustion, the victim sweats heavily; presents pale, moist, cool skin; and complains of a headache, weakness, dizziness, and/or loss of appetite. A victim may also experience heat cramps, nausea, vomiting, an urge to defecate, chills, rapid breathing, confusion, and a tingling sensation in the hands or feet. To treat, pour water over the victim and fan him to speed up the coolant effect of evaporation, have the victim drink freshwater, and attempt to provide shade.

With heat stroke, the victim stops sweating, which results in red, flushed, hot, dry skin. The victim may first experience headaches, dizziness, nausea, fast pulse and respiration, and/or seizures and mental confusion. The victim can become unconscious and die if not treated quickly. To treat, create improvised shade for the victim. Have him drink water and elevate both legs if possible.

Burns

Apply a field dressing (or the cleanest material available) to the burn. Give sips of water to a casualty who is conscious and not nauseated. When treating burn casualties—

- DO NOT remove clothing stuck to the burns.
- DO NOT break any blisters.
- DO NOT apply grease or ointments to the burns.

For electrical burns, check for both an entry and exit burn from the passage of electricity through the body. An exit burn may appear on any area of the body, not necessarily opposite the entry burn.

For burns caused by wet or dry chemicals, flush the burns with large amounts of water and cover with a dry dressing.

For burns caused by white phosphorus, flush the area with water, then cover with a wet material, dressing, or mud to exclude air and keep the white phosphorus particles from burning.

For laser burns, apply a field dressing.

Common Medical Problems Associated with Sea Survival

Seasickness

Seasickness is the nausea and vomiting caused by the bobbing motion created by the wave action of a flotation device. Seasickness can result in—

- Dehydration and exhaustion.
- A loss of the will to survive.
- Others becoming seasick.
- Unclean conditions.

To treat seasickness—

- Wash both the Marine and the flotation device to remove the sight and odor of vomit.
- Keep the Marine from eating food until the nausea is gone.
- Have the Marine lie down and rest.
- Give the Marine seasickness pills if available. If the Marine is unable to take them orally, the pills should be inserted rectally for absorption by the body.

Saltwater Sores

Saltwater sores occur when skin that has abrasions or is cut is exposed to saltwater. The sores may form scabs and pus. Do not open or drain. Flush the sores with freshwater, if available, and allow to dry. Apply antiseptic, if available.

Blindness/Headache

Irritants or the effects of the sun's rays reflecting off the water can cause temporary blindness or headaches. If flames, smoke, or other irritants get in your eyes, flush the eyes immediately with saltwater, then with freshwater, if available. Apply an ointment, if available. Bandage both eyes for 18 to 24 hours or longer if the damage is severe. If glare from the sky and water causes your eyes to become bloodshot and inflamed, bandage the eyes lightly. Try to prevent this problem by wearing sunglasses or goggles with a sunglass insert.

Constipation

This condition is a common problem associated with dehydration. For constipation, do not take a laxative if it is available; this causes further dehydration. Drink freshwater, if available.

Sunburn and Dehydration

The sun's rays reflect at all angles off the waves of the water; therefore, sunburn and dehydration are serious problems in sea survival. Try to prevent sunburn by—

- Erecting an improvised canopy, with available floating materials, to provide shade.
- Wearing your soft cover or using a cloth, such as a handkerchief, to cover your head.
- Covering your skin with sunscreen or lip balm from your first aid kit. Your lips, nostrils, eyelids, the backs of your ears, and the skin under your chin sunburns easily. If enough sunscreen cream is available, all exposed skin should be covered.

Dehydration is caused by the loss of the body's vital fluids. Dehydration in saltwater may result from a combination of factors such as a lack of water, the effects of saltwater on skin tissue, sunburn or vomiting from seasickness, and other causes. Sleep and rest and reduced water and food intake are the best ways of enduring periods of exposure. The following measures will delay the effects of dehydration:

- DO NOT drink saltwater.
- DO NOT drink urine.
- DO NOT drink alcohol.
- DO NOT smoke.
- DO NOT EAT unless water is available.

Dangerous Marine Life

You may see many types of marine life around you, and some are more dangerous than others. Generally, sharks are your greatest danger, followed by barracudas. However, most marine life will not deliberately attack a human. The most common injuries from marine life are wounds from bites, stings, or punctures. With the exception of sharks and barracudas, most injuries are a result of either trying to catch game or from contact abrasion with marine life. To treat an injury that results from a dangerous marine animal bite—

- Pack the wound with gauze and then apply a pressure bandage, if available.
- Treat for shock.
- Obtain medical attention as soon as possible.

For a more detailed information on the treatment of casualties and dangerous marine life, see MCRP 3-02F and MCRP 3-02G.

Sharks and Barracudas

Only about 20 percent of all shark species are known to attack people. Sharks have an acute sense of smell, and the smell of blood in the water will draw them to their prey. They are also very sensitive to any abnormal vibrations in the water; therefore, the sound caused by a struggling swimmer or underwater explosions will attract them.

Barracudas are bold and inquisitive fish. They have been known to attack men who are wearing shiny objects. Barracudas may charge at lights or shiny objects at night.

A group of swimmers can maintain a 360 degree watch while in the water. Therefore, to protect yourself from sharks and barracudas, stay with other swimmers. The group members can either frighten away or fight off sharks or barracudas better than one person can. Keep all your clothing on, including footwear. Historically, sharks attack unclothed individuals in groups first, mainly in the feet. Clothing also protects you against abrasions from a shark's tough skin should the shark brush up against you. Avoid urinating heavily, let urine dissipate between discharges. If you must defecate, do so in small amounts and toss it as far away from the group as possible. Do the same if you must vomit.

If attacked, the use of firearms by swim sentries should be used with extreme caution because of the risk of injury to other swimmers. If unarmed or unable to make an improvised weapon, kick and strike the shark. Avoid using your bare hands to strike the shark, injury can result to your hands due to a shark's tough skin. Target areas on a shark are the gills, eyes, and underbelly. Blows to the snout are also not recommended because a shark will tilt its head up and thrust its jaws forward when biting.

Sea Snakes

Sea snakes are venomous and sometimes found in mid-ocean. They are unlikely to bite unless provoked. Your best protective measure is to avoid them.

Poisonous Fish

Many reef fish are toxic, and can kill you if eaten. Generally, poisons are present in all parts of the fish, but especially concentrated in the liver, intestines, and eggs.

Turtles and Moray Eels

Turtles and moray eels normally inflict minor bite wounds. Treat this type of injury by cleaning the wound.

Corals

Coral, dead or alive, can inflict painful cuts. Clean all coral cuts thoroughly. DO NOT use iodine to disinfect any coral cuts. Some coral polyps feed on iodine and may grow inside the flesh.

Jellyfish, Portuguese Man-of-War, Anemones, and Others

This group of marine animals inflicts injury by stinging their victims with their tentacles. Contact with their tentacles produces burning pain, a rash, and small hemorrhages on the skin. Shock, muscular cramping, nausea, vomiting, and respiratory distress may also occur. Gently remove the clinging tentacles with a towel and wash or treat the area. Use diluted ammonia or alcohol and talcum powder to treat the injury if available.

Spiny Fish, Urchins, Stingrays, and Cone Shells

These animals inject their venom by puncturing the skin with their spines. General signs and symptoms include swelling, nausea, vomiting, generalized cramps, diarrhea, muscular paralysis, and shock. Deaths are rare. Treatment consists of soaking wounds in hot water, if available, to deactivate heat-sensitive toxins.

Chapter 4
Negotiating Water Obstacles

Marines face water obstacles in saltwater, freshwater, and brackish water (where freshwater and saltwater meet). These water environments differ considerably and pose distinct problems for Marine tactical units and swimmers. Saltwater obstacles include tides, surfs, and currents. Brackish water obstacles include back bays. Freshwater obstacles include rivers and canals.

Tides

Tides are periodic changes in the surface levels of oceans, bays, gulfs, inlets, and rivers. The Moon's and Sun's gravitational pulls cause tides. Depending on the situation, tides can either help or hinder Marines in their endeavors to conduct amphibious or riverine operations. Tides can provide sufficient depths to allow for the passage of landing craft or boats over reefs, trees, and other underwater obstructions. On the other hand, tides can render a river fording site unusable to tactical vehicles. Direction, level of change, and amount of change determine tidal nomenclature.

Tides that show change in direction are flood tides and ebb tides. Rising tides are known as flood tides. Falling tides are known as ebb tides.

Tides that show extreme levels of change are high tides and low tides. High tide is the period when water is at its greatest depth. Low tide is the period when water is at its most shallow depth.

Tides that show the least and most amount of change are neap tides and spring tides, respectively. Neap tides have the least amount of change in water levels between high and low tide. Neap tides occur at the half moon, when the Sun and the Moon are aligned at a 90-degree angle with the Earth. In this position, the Sun's and Moon's gravitational pulls offset each other. Spring tides have the highest floods and lowest ebbs. Spring tides occur at or shortly after the new moon or full moon when the Sun, Moon, and Earth are approximately in line. In this position, the Sun's and Moon's gravitational pulls are combined.

Surf

Waves break upon entering shallow water and create surf. The offshore area where waves break is the surf zone, which can present many hazards. Breaking waves often trap air bubbles and create a foamy appearance. Bubbles lower the water's density and decrease buoyancy. Move through foamy surf as quickly as possible. The type of wave determines your survival technique.

Wave action moves you toward shore. Lie on your back or side with your head pointing in the direction of the beach and your feet pointing into the waves.

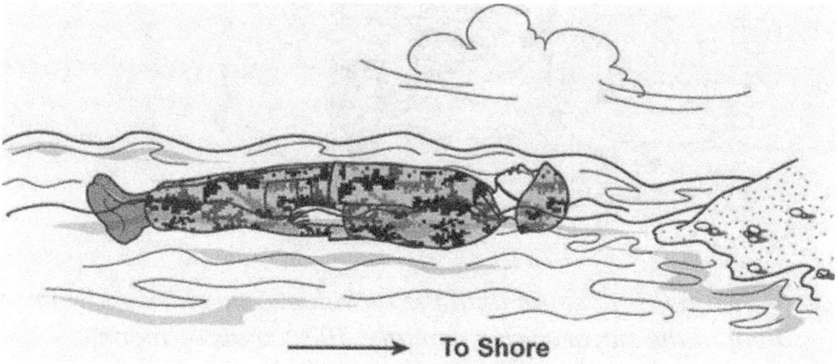

As one wave approaches the beach, another drains away; relax and do not swim against the draining water. When a new wave is within about 3 yards, start swimming toward shore. Continue to swim until the wave lifts you and moves you toward the beach. Once the wave loses forward momentum, relax and repeat the cycle. If nearing rocks, turn your body and approach feet first to reduce the chance of striking your head and arms.

Plunging Waves

A plunging wave is a breaker whose crest curls forward and falls ahead of its base. Because of its power and underwater turbulence, a plunging wave poses the greatest surf threat. If caught in a plunging wave, you can be pulled underwater and pitched about violently. This can cause you to panic, which can increase your chance of drowning. Perform the following steps to escape a plunging wave:

Tuck into a ball with your head against your knees and your forearms locked around your legs, just below the knees.

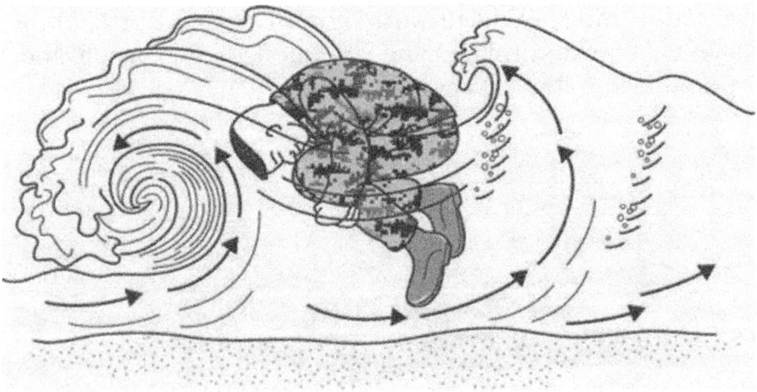

Relax in this position until the turbulence subsides and you float to the surface. This can take 30 seconds or more.

Swim toward shore.

> *Note:* If threatened by another plunging wave, dive underwater into the wave.

Spilling Waves

A spilling wave does not break. Instead, its crest slides forward without curling. A spilling wave creates less turbulence and poses less of a threat than a plunging wave. If caught in a spilling wave, relax and let the wave carry you to shore.

Surging Waves

A surging wave occurs on a beach with a steep underwater gradient. It never really breaks, but the crest rises while the base slides up the beach with great force and speed. Once the wave reaches its highest point on the beach, it rushes back as quickly as it surged forward. If you are standing on the bottom when a surging wave advances or retreats, the wave can knock you off your feet and pull you into the surf zone. Do not try to stand or walk on the bottom. Swim toward the beach as soon as possible.

Currents

Offshore Currents

An offshore current occurs outside the surf zone. Typically, it occurs at bay entrances, in island channels, and between islands and the mainland. An offshore current flows parallel to or away from shore. If the offshore current is created by tides, its current strength and direction vary at different times of the day.

If caught in an offshore current, you may be carried in a direction you do not want to go. DO NOT try to swim directly to safety. If the current is moving directly away from the shore, relax and wait until the current dies out or turns toward land. Once the current subsides, use any survival stroke to swim toward shore. If the current is moving parallel to shore, use any stroke to move at an angle across the current and toward shore.

Rip Currents

A rip current occurs when waves pile water against the shore faster than the water can drain. The water flows rapidly along the beach until it is deflected seaward by a bottom obstruction. Then the rip current flows through the surf zone and into open water at a speed of up to 2 knots. This action can cut deep trenches in the sand. A rip current dies out once it reaches open water (usually within a few hundred yards of the shore).

A rip current can pull you out to the open sea. If caught in a rip current, DO NOT try to swim against the current. A rip current moves faster than most people swim, and it is impossible to swim to shore once caught in it. Relax and stay afloat until the current runs out. Once the current subsides, use a survival stroke to move parallel to the shore until you are out of the current, then begin swimming toward shore.

Littoral Currents

A littoral current occurs when a wave breaks against a beach at an angle. This current flows parallel to the shoreline and does not pose a great threat. If caught in a littoral current, use the combat travel stroke to swim across it at an oblique angle.

Back Bays

Once on the beach, you often face one or more rows of low hills called dunes. Behind the dunes, you may encounter a low-lying stretch of ground thickly covered with scrub trees and bushes. This area gives way to wetlands known as back bays. Back bays consist of muddy islands that are almost submerged during flood tides and separated by channels of brackish water of varying depths. Channel bottoms usually contain soft mud. Back bays pose major obstacles to vehicular traffic.

Infantry can cross back bays, but only with great effort. If crossing back bays by foot, consult detailed navigation charts and use the following guidelines to plan your route:

- Avoid water less than waist deep; walking in shallow water or soft mud is extremely tiring.
- Avoid back bay islands; these low-lying islands are usually too muddy to support foot traffic.
- Seek out deep water; floating with a pack is less tiring than walking through shallow water or soft mud.
- Seek out sand, shell, gravel, or stone bottoms; these firmer bottoms generally ease travel and help conserve energy.

Rivers and Canals

A river is a large, natural stream of water that empties into a larger body of water. The slope of the riverbed and the volume of water in the river determine its current.

Canals resemble small rivers or streams in their width and depth, but usually lack any significant current. Climbing out of these waterways can be difficult if the canal is flanked by steep banks.

Chapter 5
Fording Waterways

WARNING

Fords are dangerous. Cross as quickly as possible.

A ford is any site in a river, stream, or canal where the water is shallow enough for troops or vehicles to cross without using flotation devices. Canal bottoms are usually too soft to support fording vehicles, and wading infantry frequently stumble.

The tactical situation dictates the location of the fording site. Seek fords that are protected from enemy observation and that allow for adequate supporting fires. A night fording takes at least one and a half times as long as a daylight fording.

NOTE: In this chapter, the term "river" refers to rivers, streams, and canals.

Selection of a Ford Site

The following table identifies desirable fording characteristics:

Characteristics	Comments
Concealment	The ford hides personnel and vehicle movement from enemy observation.
Accessibility	The ford should have low banks with gentle gradients. This allows a free flow of traffic at both the entrance and the exit.

Characteristics (Continued)	Comments (Continued)
Slow Current	The ford's current should not exceed 1.5 meters per second if possible.
Firm Footing	The ford's bottom, entry, and exit composition should be firm enough to support traffic. Do not drive a vehicle over any bottom composition that a 2-inch diameter stick can be pressed into more than 1 or 2 inches.
Gently Sloped Channels	The ford's entry and exit points should be gently sloped. If possible, locate a portion of the stream where the channel is not actively shifting.
Depth	The fording depth is less than or equal to the least capable vehicle.

Determine the Slope

Units move into and out of water faster and more quietly if entry and exit points are not steep or muddy. Slope is the amount of change in ground horizontal distance (run) and in vertical elevation (rise) from one point to another. Slope is usually expressed as a percentage. You can use a clinometer, map, or line of sight and pace to measure the percentage of a slope.

Clinometer

A clinometer measures percentage of a slope. It is a component of the M9 armored combat excavator, and organic to many Marine engineer units. However, engineers are also taught (and often use) field-expedient methods to determine slopes and gradients.

Map

A map measures the horizontal distance along a desired path. Determine the difference in elevation between the path's starting and ending points. From a map's scale, you can determine the distance between two points. From a map's elevation lines, you can determine the difference in height between the same two points. Both figures must be in the same unit of measure (e.g., feet, meter). Divide the elevation (rise) by the distance (run) and multiply by 100.

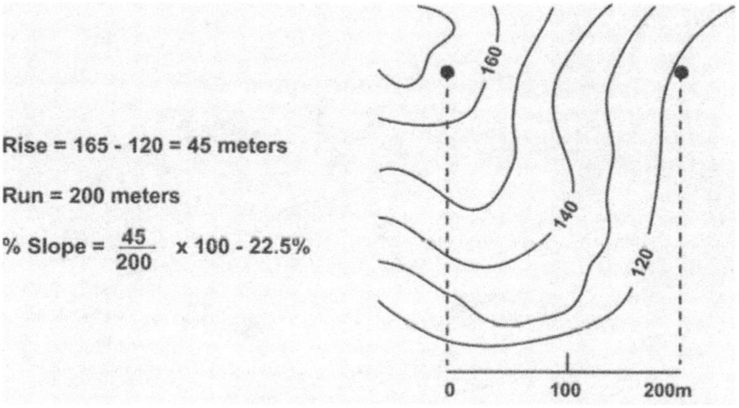

Line of Sight and Pace

To determine line of site and pace, stand at the bottom of the slope, keep your eyes level, pick a spot on the slope, then pace the distance. The number of paces multiplied by a standard measure of 0.75 meter determines the run. The eye-level height (usually 1.5 to 1.7 meters) determines the rise. Repeat this procedure until you have covered the entire distance you want to measure for each spot (vertical and horizontal). Add the vertical distances to provide total rise and the horizontal distances to provide total run.

Determine the Current Speed

Current speed increases as channels narrow. It may be necessary to locate a wider ford location to obtain a slower stream current. The following steps are used to calculate the speed of the current:

Determine points A and B along a channel. Then measure the distance between those two points.

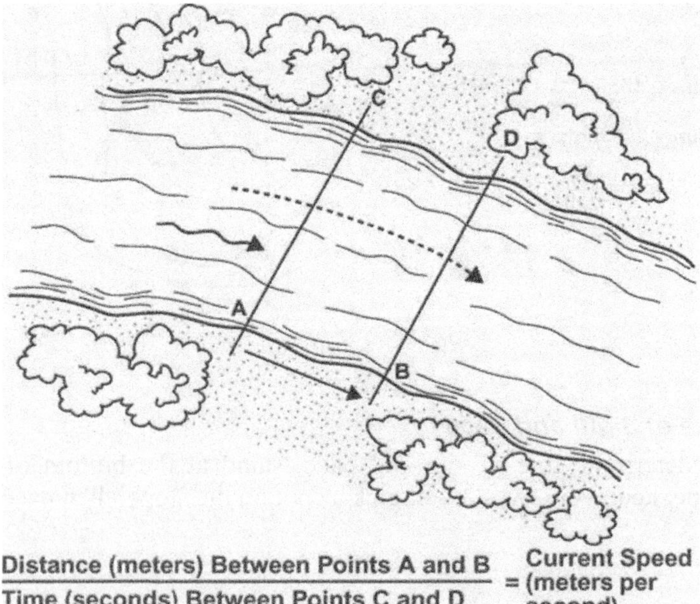

$$\frac{\text{Distance (meters) Between Points A and B}}{\text{Time (seconds) Between Points C and D}} = \text{Current Speed (meters per second)}$$

Sight directly across the water from points A and B to locate points C and D.

Throw a floating object (e.g., a stick) upstream from points A and C. Observe the object as it floats toward points B and D.

Subtract to find the time it takes for the object to float from start to finish.

Do not attempt to swim across currents that are moving faster than 1.5 meters per second. Equivalents of this speed include—

- Quick-time march rate of 120 counts per minute with one, 30-inch step at each count.
- 5 feet per second.
- 3.5 miles per hour.
- 5.5 kilometers per hour.

Measure River Width

A river's width can be estimated from the width of its symbol on a scaled topographic map. If this is not possible, use the following compass techniques:

Stand at the water line (A).

Shoot an azimuth to a point on the opposite bank (B).

Move upstream or downstream until you are at a point (C) where you can shoot an azimuth 45 degrees larger or smaller than the original azimuth.

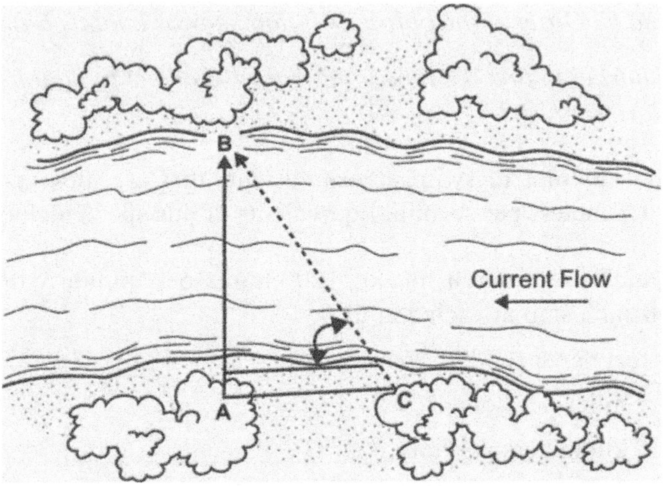

Measure the distance between points A and C. The distance calculated equals the river's width.

Calculate Downstream Drift

A river's current causes personnel and equipment to drift downstream. If personnel and equipment are aimed straight across the river, they will sideslip downstream as they move across the current to the other shore. Therefore, personnel and equipment crossings must compensate for the effects of a river's current and entries are usually made upstream of the desired exit point. Use the formula below to calculate downstream drift.

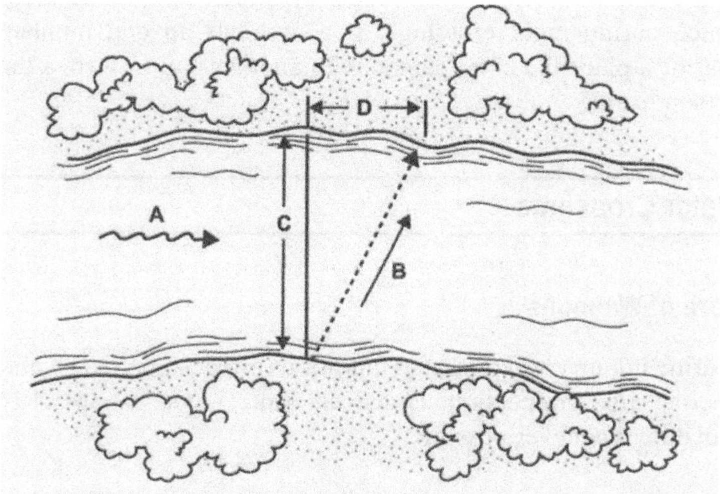

$$\frac{\text{Current Speed (A)}}{\text{Crossing Speed (B)}} \times \text{River Width (C)} = \text{Downstream Drift (D)}$$

NOTE: The crossing speed for a swimmer across a river may vary but is generally limited to 1 meter per second. All measurements must be in the same unit of measure (e.g., meters, feet).

The Buddy System

Whenever a Marine unit must enter into or operate on the water, a "buddy system" is employed in which every Marine is paired with a swimming partner. The buddy system matches an experienced swimmer with a weak swimmer. The experienced swimmer assists and encourages the weaker swimmer and bolsters confidence during night crossings. If a unit has an odd number of Marines, place the extra person with another pair to form a three-person team.

Water Crossings

Care of Weapons

Marine infantry weapons and munitions are designed to be able to operate after immersion. However, protect your weapons from moisture whenever possible.

A gas-operated weapon can malfunction if water travels down the barrel and enters the gas tube. To protect the gas tube—

Close the weapon's bolt before entering the water.

Seal the muzzle with a condom, balloon, plastic spoon wrapper, or other form of waterproof material.

Tie or melt the protective cover to create a watertight seal.

When the muzzle's protective cover is no longer needed, remove it. Open the bolt and inspect the barrel. If the tactical situation permits, swab excess moisture from the barrel. Test fire automatic weapons, if possible. Field strip and clean weapons as soon as possible. If time does not allow for a complete inspection, rinse inaccessible areas with small amounts of diesel fuel, then dry.

ISOMAT Raft

Construction of an ISOMAT raft is time-consuming. This type of raft should not be employed as part of an attack, but used for logistical purposes (e.g., evacuating stretcher cases, transporting supplies). Use the following steps to build an ISOMAT raft:

Wrap ISOMAT sleeping pads around sturdy sticks.

Use parachute cord and square knots to tie the pads securely in place and to lash stick ends together in a rectangle.

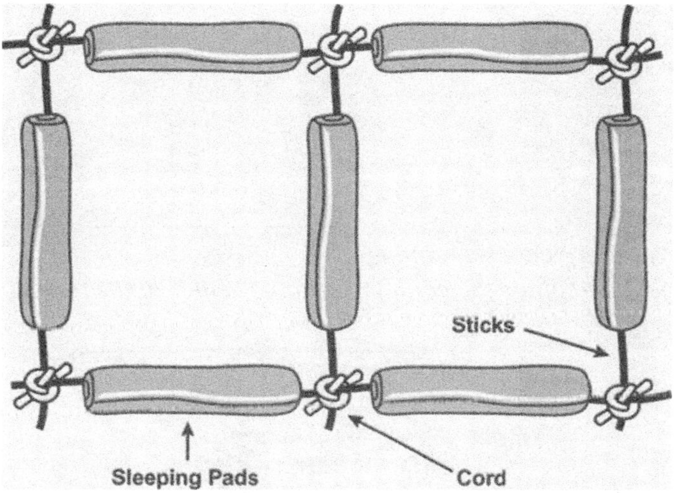

Note: The ISOMAT raft pictured can support several hundred pounds. However, the cargo will get wet if not properly waterproofed.

Poncho Raft

A poncho raft can support two Marines and their equipment and is well suited for long crossings. Use the following steps to build a poncho raft:

Inspect two ponchos and ensure they are serviceable.

Lay one poncho flat on the ground, with the hood-side up.

Cinch the hood tightly to form a gooseneck or tie in a knot.

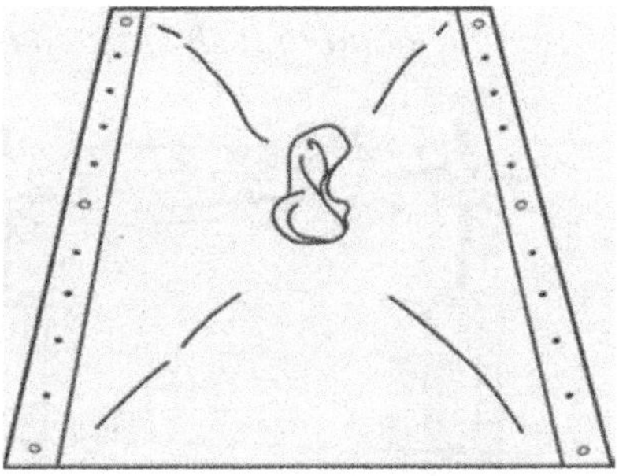

Pad sharp edges of equipment and place the equipment in the center of the poncho.

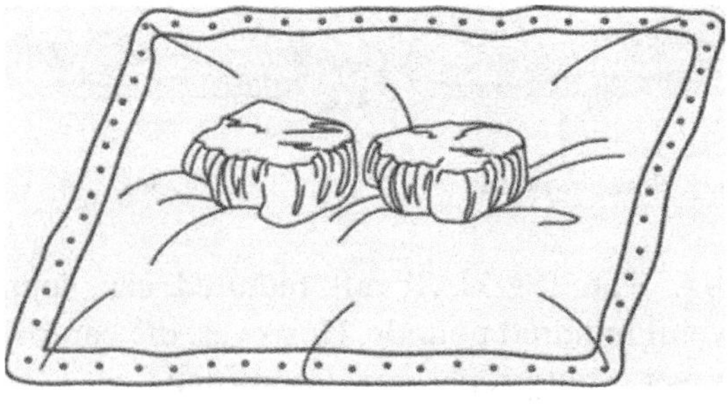

Place the second poncho over the equipment, rubber side up, and hood facing down.

Snap the edges of the two ponchos together.

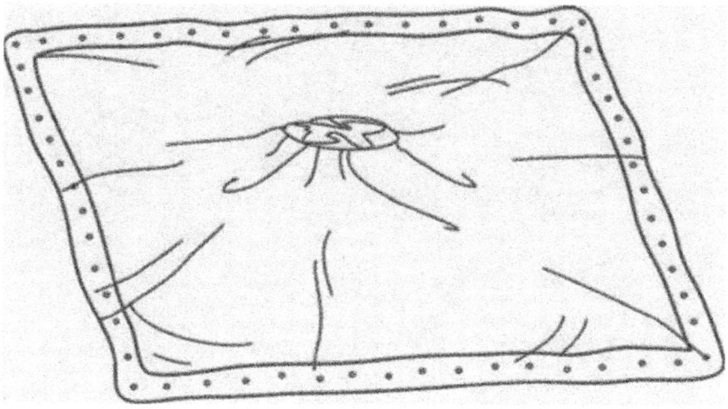

Roll the edges toward the equipment.

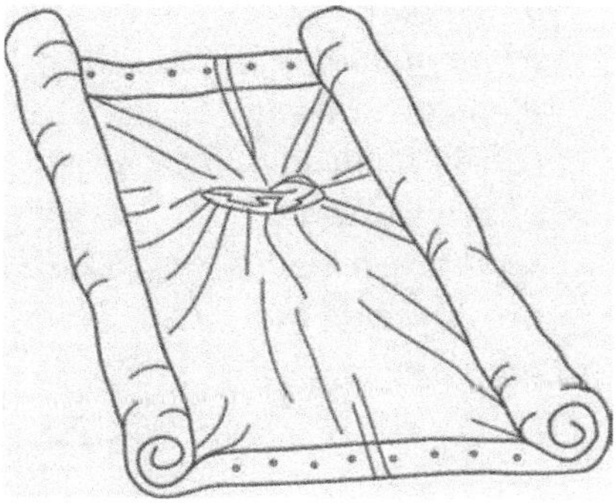

Roll the edges into pigtails and tie them off.

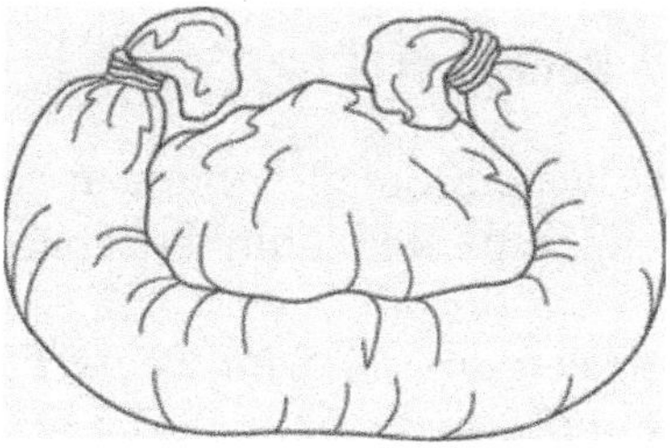

Pull the pigtails together over the top and lash them securely.

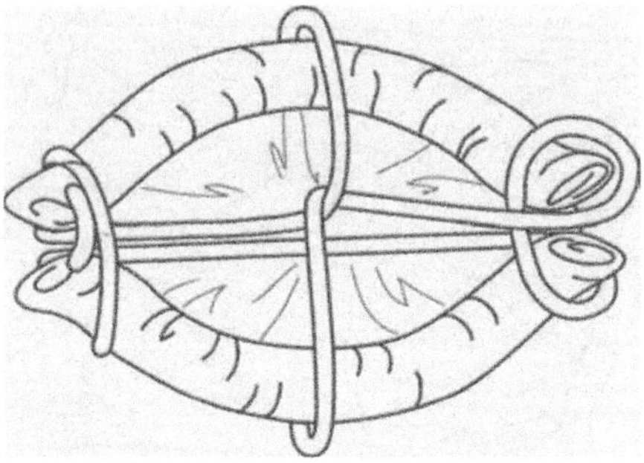

Protect the raft from brush punctures while placing it in the water. Swim across the water obstacle while security elements are covering the far shore.

Construction of a Pack Raft

You will need two waterproofed packs and two M16A2 service rifles. The following steps are required to construct a pack raft:

Place two packs side-by-side with the pack frames on the deck. The tops of the packs are opposite of each other.

Loosen the main compartment straps on both packs.

Insert one rifle on each end between the straps and the packs. The muzzles are opposite of each other. The rifles serve as one means to secure the packs together. Place the front sight post under the top flap.

Tighten the straps so that the rifles and packs are secure.

Take the excess strap on the inner side of each pack and secure it to the opposite pack to better secure the two packs together.

Take the excess straps on the outer sides of the packs and use those straps as safety lashing for the rifles.

Tuck the excess straps and check to make sure the rifles and packs are secure.

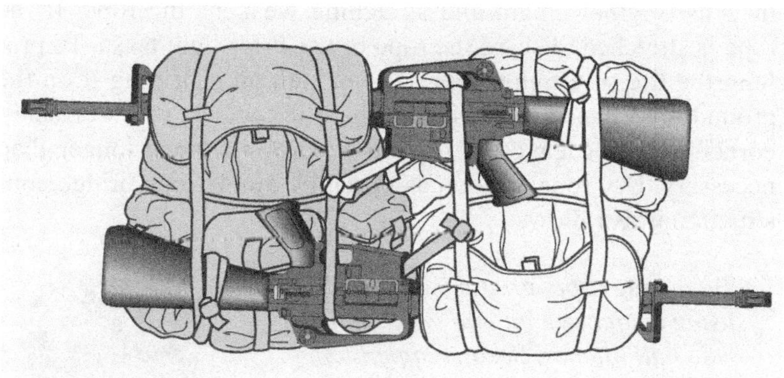

Single-Rope Bridge

A single-rope bridge offers a temporary and quick way to cross small rivers. It also provides extra security while crossing swift waters. At night, it prevents straggling, and guides units precisely from one side of the river to the other side. If crossing a river at night, plan for at least one single-rope bridge.

If your unit is crossing a river with swift currents or water depths above 4 feet, the unit is carrying sufficient rope to span the crossing site, and the tactical situation permits, secure the rope on near and far banks to provide a hand-hold for crossing Marines. This reduces the time required for the entire unit to cross and provides a degree of comfort/confidence for poor swimmers. Use a squad-sized bridge team to construct a single-rope bridge. Station several strong swimmers at the water's edge to help anyone who has trouble crossing.

Nylon rope is normally coiled in 120 foot lengths. It is 0.6 inches in diameter and has a breaking strength of about 3,840 pounds. Over time, a nylon rope can stretch to as much as one-third more than its original length and stretching weakens the rope. If the rope is stretched, discard the rope or use it for light tasks. To prolong the life of a nylon rope, do not step on it or drag it on the ground. Pad the rope in places where it contacts rocks or sharp corners. Do not leave the rope knotted or stretched longer than necessary. Dry rope as soon as possible. Single-rope bridge construction is as follows:

Tie a sling rope around your waist using a square knot and two, separate half hitches. See the appendix for detailed information on knots.

Attach a locking steel carabiner to the sling rope.

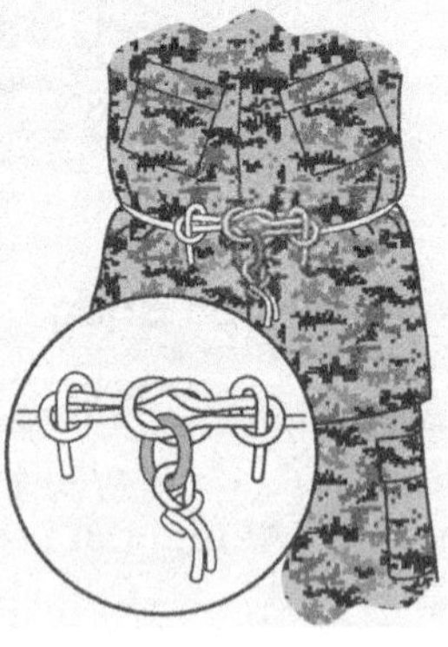

Tie a bowline knot in the running end of the bridge rope and attach it to the carabiner.

Temporarily secure the other end of the rope to a tree on the near shore.

Enter and cross the water.

NOTE: Carry only your weapon and ammunition.

Exit the water on the opposite shore.

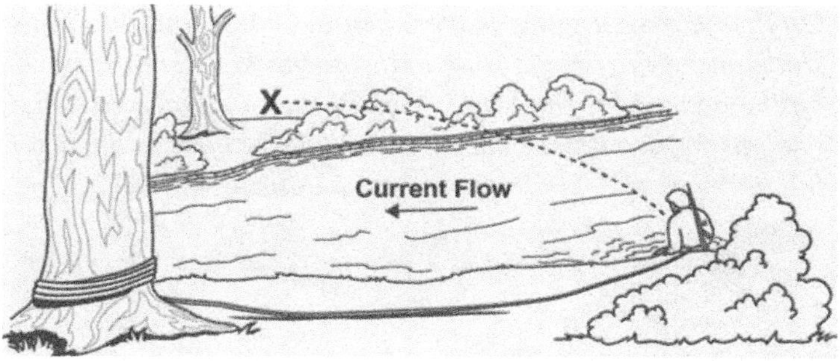

Prepare your weapon for use. Unhook the bridge rope from the carabiner at your waist, and tie the bridge rope to a sturdy tree using a round turn and two half hitches.

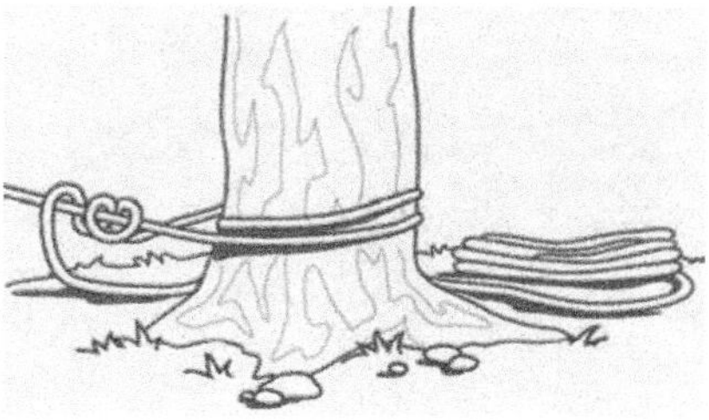

Conduct a box reconnaissance of the opposite shore.

On the near shore, have another Marine prepare to tighten the rope. That Marine should place a transport tightening system in the bridge rope by tying a double butterfly knot and placing two carabiners in the butterfly. See the appendix for detailed information on knots.

The Marine should pass the running end of the bridge rope around the downstream side of the near shore anchor point and through the two carabiners.

Pull the butterfly knot approximately one-third of the distance across the river.

Secure the bridge rope to an anchor point using a round turn and two half-hitches.

On the near shore, the Marine helping you should pull the slack out of the bridge rope until the butterfly knot is back on the near side. The bridge rope is then tied off against itself using two half hitches with a quick release in the last half hitch.

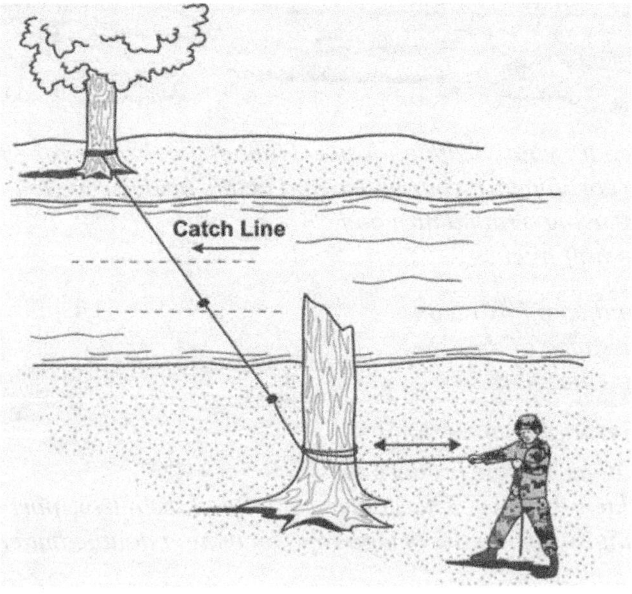

Note: The single-rope bridge must be as tight as possible so it will not sag when used.

If you lose your footing and fall into the water, swim with the current to the closest shore. Swimming against the current is dangerous and quickly causes fatigue.

High and Dry Crossings

If the single-rope bridge is high enough, suspend yourself below the single-rope bridge and above the water. Use the following steps to suspend yourself from a single-rope bridge and then pull yourself across the water:

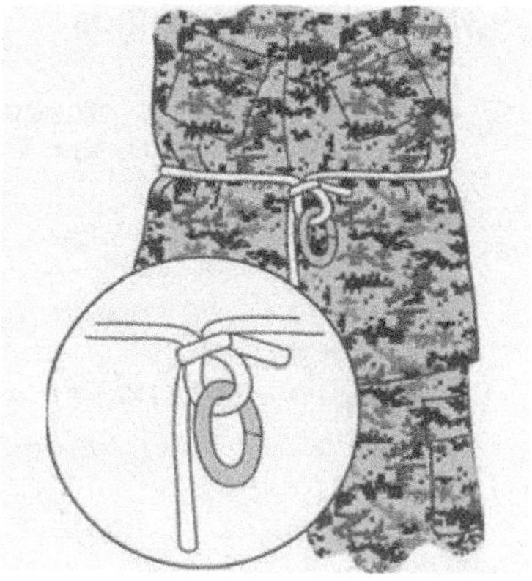

Tie a sling rope around your waist using a bowline. Ensure that the knot is tight.

Attach a carabiner through the bowline's loop. The carabiner's gate faces up.

Secure your helmet chin strap.

Face the single-rope bridge with your left shoulder toward the far shore.

Grasp the bridge rope in both hands.

Swing your body beneath the single-rope bridge with your head toward the far shore. Cross your ankles above the bridge rope.

Arch back until the carabiner contacts the bridge rope. Connect the carabiner to the bridge rope. Allow the carabiner to bear your body's weight.

Pull yourself across the single-rope bridge, hand over hand, to the far shore.

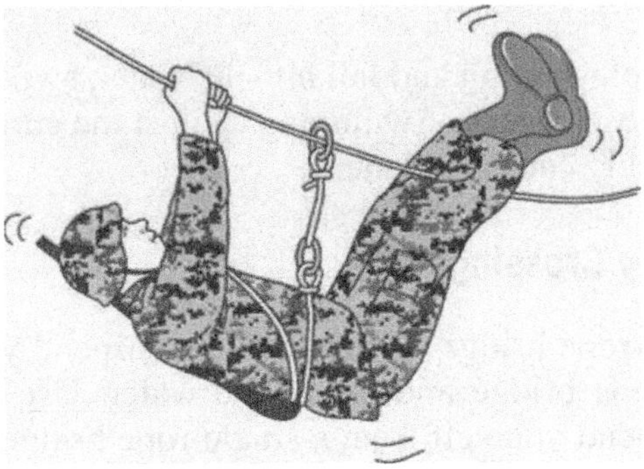

Swift Current Crossings

A single-rope bridge prevents being knocked down and swept away by a swift current. Use the following steps to move through a swift current:

Tie one end of a sling rope around your waist using a bowline.

Tie the running end of the sling rope in another bowline, and attach a carabiner to the bowline's loop.

Step up to the bridge. Face upstream.

Hook the carabiner to the single-rope bridge.

Walk sideways into the river while grasping the bridge rope in both hands.

Use the single-rope bridge for balance and remain standing, if possible.

Continue to move sideways through the river to the far shore.

Slow Current Crossings

If you face little or no current, it is not necessary to hook up to a bridge rope with a carabiner. Lie on your back in the water beneath the single-rope bridge. Support your body weight with your waterproof pack. Use the bridge rope and pull yourself hand over hand across the river.

Removal

If you are the last Marine waiting to cross, pull on the standing end of the rope to release the knot, then tie the rope around your waist using a bowline. The Marines on the far shore will pull you through the water.

Appendix
Knot Tying

Square Knot

The square knot is used to secure two ropes of equal diameter together so they form one continuous rope that will not slip.

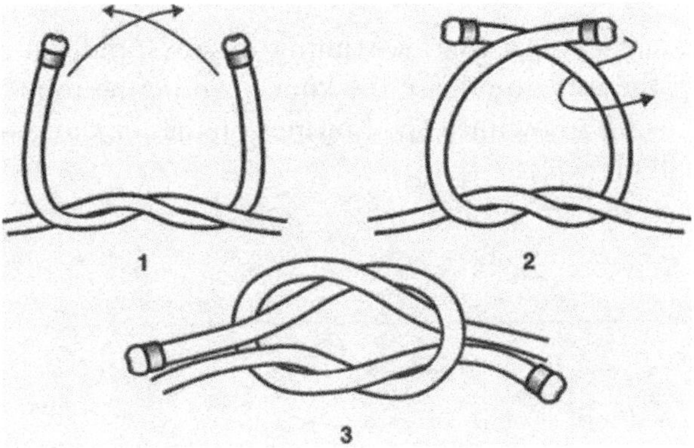

Bowline

The bowline forms a loop that will not tighten or slip under strain. It is easily untied.

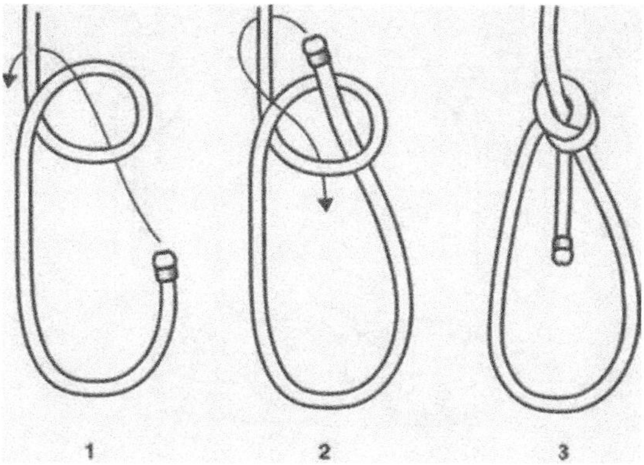

Hitches

Hitches are used to dress (prepare) knots and secure loose ends.

Half Hitch

The half hitch is used to tie a rope to a tree or to a larger rope. It will hold against a steady pull, but is not a secure hitch. It is frequently used to secure the free end of a rope.

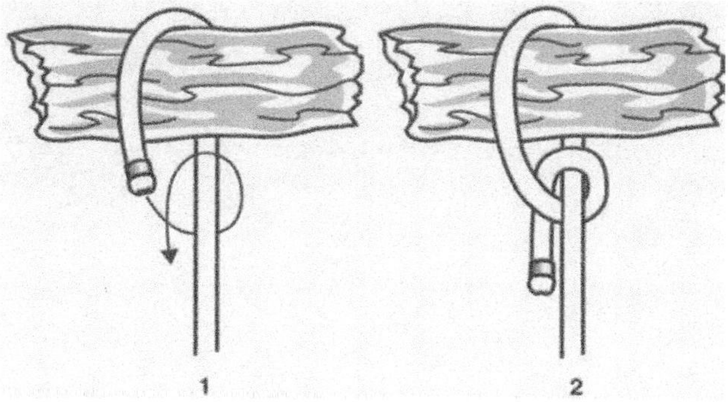

Two Half Hitches

Two half hitches can be used to secure the running end of a rope.

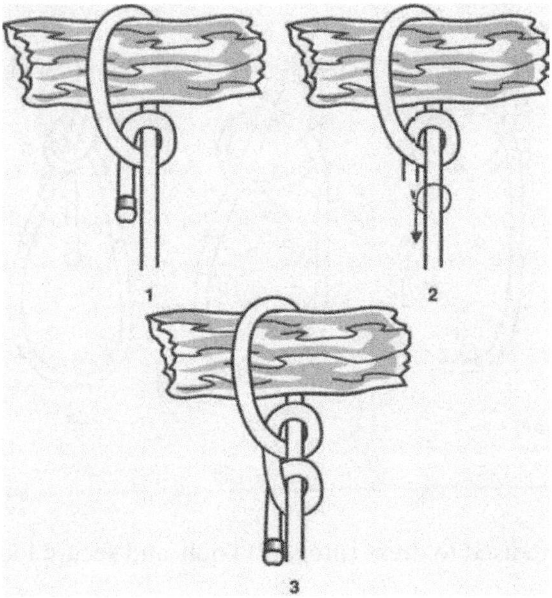

Round Turn and Two Half Hitches

A round turn and two half hitches can be used to fasten a rope to a tree. This hitch does not jam.

Butterfly Knot

The butterfly knot is used for anchor lines. Tied properly, it will not tighten on itself to the point that it cannot be easily untied. The butterfly knot can be used to pull a rope bridge taut. This knot can be used to tighten a fixed rope when mechanical means are not available. It will not jam if a stick is placed between the two upper loops.

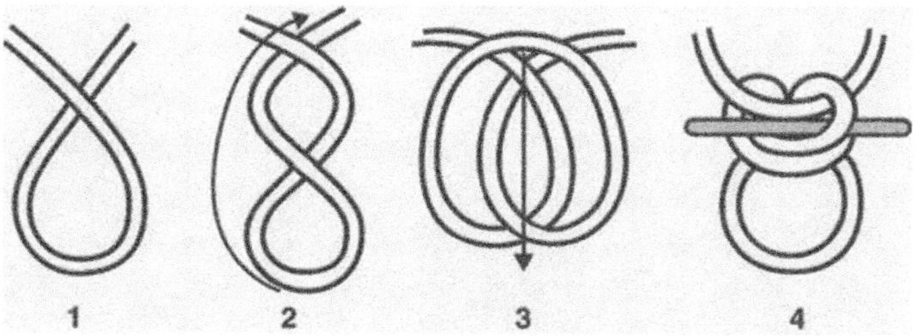

Training Circular
No. 90-1

***TC 90-1**

Headquarters
Department of the Army
Washington, DC, 19 May 2008

Training for Urban Operations

Contents

Page

PREFACE .... **vi**

Chapter 1 INTRODUCTION **1-1**

Section I. DEFINITIONS **1-1**
Military Operations on Urbanized Terrain 1-1
Urban Operations 1-1

Section II. TRAINING STRATEGY **1-2**
Training Phases 1-3
Training Environments 1-4
Mission-Essential Task List 1-7
Battalion Integrated Training Plan 1-7
Risk Assessment 1-7

Section III. TRAINING FACILITIES **1-9**
Urban Assault Course 1-9
Shoot House 1-10
Breach Facility 1-10

Section IV. CONDUCT OF TRAINING **1-11**
Crawl 1-11
Walk 1-11
Run 1-11
After-Action Review 1-11

Section V. PREREQUISITE TRAINING **1-12**
Command and Staff Training 1-12
Leader Training 1-12
Individual Training 1-12
Individual Task Training 1-13
Collective Task Training 1-14

Chapter 2 URBAN ASSAULT COURSE **2-1**
Purpose 2-2
Station 1: Individual and Team Task/Technique Trainer 2-5
Station 2: Squad and Platoon Task/Technique Trainer 2-11
Station 3: Grenadier Gunnery Trainer 2-33
Station 4: Urban Offense/Defense Building, Task/Technique Trainer 2-39
Station 5: Underground Trainer 2-45

Distribution Restriction: Approved for public release; distribution is unlimited.

*This publication supersedes TC 90-1, 1 April 2002.

Chapter 3 SHOOT HOUSE 3-1
Purpose 3-1
Description 3-1
Training Audience 3-3
Training Information 3-4
Conduct of Training 3-4
Training Support Requirements 3-14
Training and Evaluation 3-15

Chapter 4 BREACH FACILITY 4-1
Purpose 4-1
Description 4-2
Training Audience 4-3
Training Information 4-4
Conduct of Training 4-5
Training Support Requirements 4-6
Training and Evaluation 4-7
Safety Considerations 4-7

Chapter 5 COMBINED ARMS COLLECTIVE TRAINING FACILITY 5-1
Purpose 5-1
Description 5-1
Tactical Exercise Without Troops 5-6
Collective Training Exercises 5-6
Infantry Rifle Platoon STX Attack 5-7
Mechanized Infantry Platoon STX Defend 5-12
Infantry Rifle Company FTX Attack 5-20
Mechanized Infantry Company or Team FTX Defend 5-29
Targetry 5-41
Considerations for Training with Armor 5-42
Additional Considerations 5-45

Appendix A TRAINING TIPS A-1
Concurrent Training A-1
Demonstrations A-2
Exercises A-2
Evaluators A-5
Tactical Engagement System A-5
Navigation A-5
Limited Visibility Training A-6
Safety and Ammunition Considerations A-6
Opposing Forces A-7
Precision Marksmanship Training A-7
Physical Readiness Training A-8
Varied Scenarios A-8
Building Facade Walls, Windows, and Doors A-8
Use of Older or Nonstandard Facilities A-9

Appendix B UNIT TRAINING MATRIX B-1

Appendix C STABILITY AND SUPPORT OPERATIONS UNIT TRAINING MATRIX C-1

Appendix D LEADER DEVELOPMENT TRAINING D-1
Simulations Training D-1
Phases of Training D-3
Example Leader's Course POI D-5
Realistic Training off Federal Facilities D-5

GLOSSARY Glossary-1
REFERENCES References-1

INDEX Index-1

Figures

Figure 1-1. Urban Operations Training Support Strategy. 1-2
Figure 1-2. Combined Arms Collective Training Facility. 1-6
Figure 1-3. Active Army, Home Station, Urban Operations Training Strategy 1-8
Figure 1-4. Army National Guard Urban Operations Training Strategy. 1-8
Figure 1-5. Urban Assault Course 1-9
Figure 1-6. Shoot House. 1-10
Figure 1-7. Breach Facility. 1-10
Figure 2-1. Urban Assault Course 2-2
Figure 2-2. Individual and team task/technique trainer, Station 1. 2-5
Figure 2-3. Example Station 1 scenario, *Varied Entry Positions*. 2-8
Figure 2-4. Example Station 1 scenario, *Multiple Room-Clearing Techniques*. 2-9
Figure 2-5. Squad and platoon task and technique trainer, Station 2. 2-11
Figure 2-6. Conducting tactical movement. 2-14
Figure 2-7. Team B--Clearing a building. 2-15
Figure 2-8. Team moving across the hall. 2-15
Figure 2-9. Team B--Moving in cross cover formation. 2-16
Figure 2-10. Team A--Clearing a hallway intersection. 2-16
Figure 2-11. Team B--Clearing a room. 2-17
Figure 2-12. Team A--Clearing room by room. 2-17
Figure 2-13. Team B--Clearing the next room. 2-18
Figure 2-14. Securing the stairs and hallway. 2-18
Figure 2-15. Preparing to clear more stairs 2-19
Figure 2-16. Team A--Clearing stairs and landing. 2-20
Figure 2-17. Preparing to gain a foothold 2-21
Figure 2-18. Securing and moving noncombatants to the command post. 2-22
Figure 2-19. Executing breach. 2-23
Figure 2-20. Moving to the foothold. 2-24
Figure 2-21. Breaching the wall. 2-25
Figure 2-22. Securing a wall in an alley. 2-26
Figure 2-23. Obtaining obscurity. 2-27
Figure 2-24. Clearing the wall. 2-28
Figure 2-25. Securing a room. 2-29
Figure 2-26. Reserve assault team moving forward 2-30
Figure 2-27. Team B, second squad-- moving into the supporting fire position. 2-31
Figure 2-28. Consolidation and reorganization. 2-32
Figure 2-29. Part 1, Grenadier Gunnery Trainer, Station 3, wooden facade. 2-34
Figure 2-30. Part 2, Grenadier Gunnery Trainer, Station 3, target engagement lane and firing positions. 2-35
Figure 2-31. Urban offense/defense building, Station 4. 2-39
Figure 2-32. Event 1, *Clear a Building*. 2-43
Figure 2-33. Event 2, *Defend a Building*. 2-43
Figure 2-34. Underground Trainer, Station 5. 2-45
Figure 3-1. Shoot House. 3-2
Figure 3-2. Command and control building and AAR room. 3-3
Figure 3-3. Event 1, *Squad Scenario (Crawl)*, force-on-targetry. 3-6
Figure 3-4. Event 2, *Squad Scenario (Walk)*, force-on-force. 3-7
Figure 3-5. Event 3, *Squad Scenario (Run)*, force-on-targetry. 3-8
Figure 3-6. Example platoon tactical scenario (*Run*). 3-9
Figure 3-7. Example company OPORD. 3-10

Figure 4-1. Breach facility. 4-1
Figure 4-2. Station 1—Door breaching....4-2
Figure 4-3. Station 2—Window breaching....4-3
Figure 4-4. Station 3—Wall breaching. 4-4
Figure 5-1. Example CACTF schematic....5-3
Figure 5-2. Platoon ODSS STX sectors....5-4
Figure 5-3. Company/team STX sectors. 5-5
Figure 5-4. Attack STX scenario. 5-8
Figure 5-5. Example FRAGO for STX Attack. 5-10
Figure 5-6. Defend STX scenario....5-13
Figure 5-7. Example OPORD for STX Defend. 5-15
Figure 5-8. Attack FTX scenario. 5-21
Figure 5-9. Example battalion OPORD for FTX Attack. 5-23
Figure 5-10. Example FRAGO for FTX Attack. 5-27
Figure 5-11. Graphic scenario of task performance for FTX Defend....5-30
Figure 5-12. Example battalion OPORD for mechanized company or team STX Defend. 5-32
Figure 5-13. Example FRAGO for FTX Defend....5-39
Figure 5-14. Urban battlespace....5-42
Figure 5-15. Example positions for Infantry riding on a tank. 5-43
Figure 5-16. Precautions for Infantrymen riding on a tank. 5-44
Figure 5-17. Actions on contact for Infantrymen riding on a tank. 5-44
Figure 5-18. Danger areas around a tank when firing the 120-mm main gun. 5-45
Figure 5-19. Tank weapon dead space at street level....5-47
Figure 5-20. Tank main gun and coax dead space above street level. 5-47
Figure A-1. Shoot-through walls and doors. A-9
Figure D-1. Echelons and uses. D-1
Figure D-2. Phases in the Live, Virtual, and Constructive Training Strategy. D-3

Tables

Table 1-1. Comparison of MOUT and UO....1-1
Table 1-2. Training environment/training phase matrix. 1-3
Table 2-1. Support requirements for Station 1. 2-7
Table 2-2. Sequence of events and times for Station 1. 2-7
Table 2-3. Support requirements for Station 2 2-13
Table 2-4. Sequence of events and times for Station 2. 2-14
Table 2-5. Support requirements for Station 3. 2-37
Table 2-6. Sequence of events and times for Station 3. 2-37
Table 2-7. Example target scenario (presentation order) for Station 3....2-38
Table 2-8. Ammunition requirements for Station 4....2-42
Table 2-9. Sequence of events and times for Station 4. 2-42
Table 2-10. Ammunition requirements for Station 5....2-47
Table 2-11. Sequence of events and times for Station 5. 2-47
Table 3-1. Sequence of events and times for the platoon....3-5
Table 3-2. Event 1, Squad Sequence (*Crawl*)....3-6
Table 3-3. Event 2, Squad Sequence (*Walk*)....3-7
Table 3-4. Event 3, Squad Sequence (*Run*). 3-8
Table 3-5. Event 4, Platoon Evaluation Sequence (*Run*)....3-9
Table 3-6. Consolidated support requirements. 3-14
Table 3-7. Related collective and supporting individual tasks....3-15
Table 4-1. Example platoon sequence of events. 4-5
Table 4-2. Consolidated minimum support requirements....4-6

Table 4-3. Consolidated annual support requirements. 4-6
Table 5-1. CACTF characteristics....5-1
Table 5-2. Sequence of events and times for STX Attack. 5-9
Table 5-3. T&EOs used to evaluate STX Attack....5-11
Table 5-4. Sequence of events and times for STX Defend....5-14
Table 5-5. T&EOs used to evaluate STX Defend. 5-19
Table 5-6. Supporting tasks for company FTX. 5-20
Table 5-7. Sequence of events and times for FTX Attack....5-22
Table 5-8. T&EOs used to evaluate STX Attack....5-28
Table 5-9. Supporting tasks for company FTX. 5-29
Table 5-10. Sequence of events and times for FTX Defend. 5-31
Table 5-11. Ammunition and pyrotechnics support requirements for a six-day FTX....5-40
Table 5-12. T&EOs used to evaluate STX Attack....5-41
Table 5-13. CACTF target requirements....5-41
Table A-1. Training exercise matrix....A-2
Table A-2. SESAMS protection....A-6
Table B-1. Army unit training matrix, brigade tasks....B-2
Table B-2. Army unit training matrix, battalion tasks. B-3
Table B-3. Army unit training matrix, company tasks. B-4
Table B-4. Army unit training matrix, platoon/squad tasks. B-5
Table B-5. Army unit training matrix, individual tasks/techniques....B-6
Table C-1. Stability and support operations unit training matrix, brigade tasks. C-2
Table C-2. Stability and support operations unit training matrix, battalion tasks....C-3
Table C-3. Stability and support operations unit training matrix, company tasks....C-4
Table C-4. Stability and support operations unit training matrix, platoon/squad tasks....C-5
Table D-1. Urban operations simulations crosswalk....D-2
Table D-2. POI for urban leader's train-the-trainer course....D-5

Chapter 1

Introduction

This chapter discusses definitions, training strategy, prerequisite training, individual task training, and collective task training for urban operations (UO). Military operations on urbanized terrain (MOUT) and UO differ in ways that reflect the changes in the way we engage in war today.

Section I. DEFINITIONS

This section defines and compares MOUT and UO.

Note: The terms "urban operations" and "UO" replaces all instances of "military operations on urbanized terrain" and "MOUT."

MILITARY OPERATIONS ON URBANIZED TERRAIN

1-1. In the past, Army doctrine characterized urban areas as a type of specialized terrain, like jungles and mountains, and called operations in this type of terrain *military operations on urbanized terrain* (MOUT). Doctrine emphasized tactical warfighting tasks on this terrain, which was characterized by densely situated buildings and streets, and assumed the enemy consisted of conventional (Soviet-type) forces. Defeating or destroying those forces overrode all other considerations.

URBAN OPERATIONS

1-2. The concept of UO assumes that the urbanized terrain is populated, and that the populace must be a foremost consideration. UO recognizes that the populace can help either side, and that it requires a working infrastructure. Urban operations, whether against an insurgent or unconventional enemy, thus vastly differ from MOUT (Table 1-1).

Table 1-1. Comparison of MOUT and UO.

	MOUT	*UO*
Enemy force	Conventional, Soviet type	Insurgent or unconventional
Mission	Defeat or destroy enemy above all other considerations	Defeat the enemy Gain and retain the good will of the populace
Populace	Not considered	Important consideration

1-3. Although certain techniques and tactics are common to both types of operations, UO goes beyond MOUT, focusing not only on defeating the enemy, but also on gaining and retaining the good will of the population. UO consider the complex and dynamic interactions and relationships between the urban area's key components--the terrain (natural and manmade), the population, and the supporting infrastructure--as an overlapping and interdependent system of systems.

1-4. A single urban operation may include the full range of Army operations--offensive, defensive, stability and support (Appendix C)--that may be executed, either sequentially or simultaneously (FM 3-06, Chapter 1). Furthermore, to accomplish an objective, UO demand working with sister services, US government agencies, nongovernmental organizations, and regional and international organizations.

Section II. TRAINING STRATEGY

The UO training strategy is primarily organized by environment and phase. Training environments include live, virtual, and constructive (Figure 1-1), and are based on the unit's METL. Training phases are initial, intermediate, and culminating. Table 1-2 shows the relationship between the training environments and the first three training phases. (See also Appendix D, which discusses leader training.)

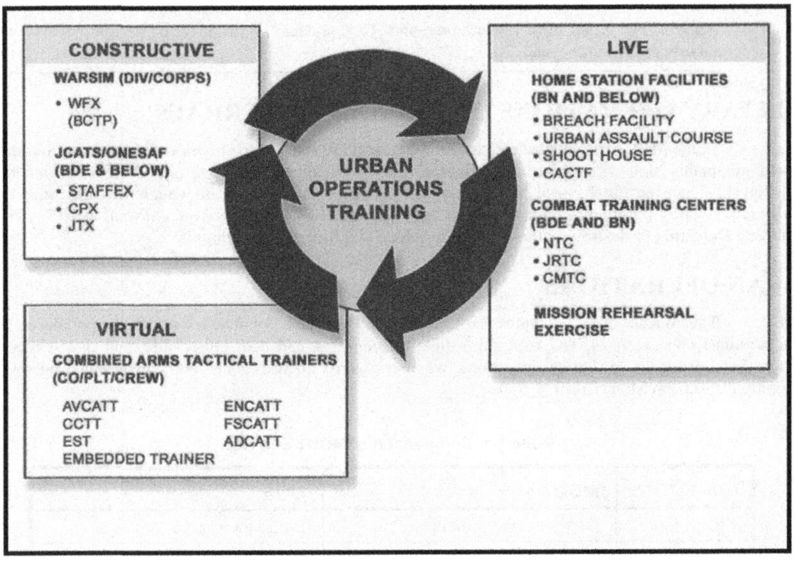

Figure 1-1. Urban Operations Training Support Strategy.

TRAINING PHASES

1-5. The three training phases are initial, intermediate, and culminating. Each of these includes some live, virtual, and constructive training.

INITIAL PHASE

1-6. This phase, which trains individual and collective tasks, is the first step in the building-block approach. It forms the basis of unit or collective training. For Soldiers and leaders, mastery of individual and crew skills is a prerequisite for UO training. The tasks train basic marksmanship skills and other Soldier tasks needed to perform any tactical operation. Included are training in a live environment (precision marksmanship) and virtual environment (Engagement Skills Trainer [EST], crew simulators, and CCTTs).

Table 1-2. Training environment/training phase matrix.

TRAINING PHASE / TRAINING ENVIRONMENT	INITIAL	INTERMED	CULMINATING
LIVE	Precision Marksmanship (Indiv / Sqd)	Breach Facility, Urban Assault Course, Shoot House (Team / Squad / Platoon)	CACTF, Platoon STX, Company STX, Battalion FTX Deployment CTC Rotation Mission Rehearsal Exercise
VIRTUAL	EST (Indiv / Sqd) Crew Simulators and CCTT (Crews Plt Ldrs, Co Cdrs)		
CONSTRUCTIVE			WARSIM (division/corps), which includes warfighting exercise (battle command training program). JCATS/OneSAF (brigade and below), staff exercises (STAFFEXs), command post exercises (CPXs), and joint training exercises (JTX)

INTERMEDIATE PHASE

1-7. This phase of training is characterized by collective task training in varied conditions such as limited visibility; different entry techniques and points; incorporation of combatants, noncombatants, and OPFOR; use of MILES; and live-fire exercises. It is the vital link between individual tasks and mission accomplishment. It should be conducted at, but is not limited to, the *Breach Facility*, *Live-Fire Shoot House*, and *UAC*. (Section III discusses these facilities in detail.) Many types of organizations can use these facilities. Trainers *must* be certified before conducting demolition training IAW local standing operating procedure (SOP) or ACOM policy. AR 385-63 prescribes safety requirements. Except for Aviation, Armor, and other branches that have no practical use for these facilities, this phase must be completed to advance to the culminating phase.

CULMINATING PHASE

1-8. The Combined Arms Collective Training Facility (CACTF, Figure 1-2) is where the culminating phase of urban operations training ends. This phase can include multiechelon, combined arms, and branch specific STX training may be conducted. Urban operations leader training occurs in leader professional development courses (Appendix D). Units have developed internal urban warfare training programs and TRADOC schools have assembled technical mobile training teams (MTTs) for marksmanship, combatives, and breaching techniques.

EDUCATION PHASE

- Urban warfare reading/video list.
- Urban operations equipment.
- Review of Infantry doctrinal publications (FM 3-21.10, FM 3-06.11, for Infantry), and corresponding publications for other branches.
- Small-unit urban movement techniques and considerations.

PLANNING AND EXECUTION PHASE

- Prepare company OPORD or WFF-related products.
- Brief platoon/company OPORD/WFF-related products to commander or staff representative.
- Participate as briefer in urban environment TEWT.

SUSTAINMENT PHASE

- Participate in company/platoon TEWT or other urban operations event.
- Execute/facilitate a company or platoon TEWT.

TRAINING ENVIRONMENTS

1-9. Both live and virtual training environments are used to train UO.

LIVE ENVIRONMENT

1-10. The live training environment consists of home station training using the Urban Assault Course (UAC), Shoot House (SH), Breach Facility (BF), Combined Arms Collective Training Facility (CACTF), Combat Training Center (CTC) rotations, and any other live training that supports the UO training.

VIRTUAL ENVIRONMENT

1-11. The virtual training environment consists of simulators such as Close Combat Tactical Trainer (CCTT) or Engagement Skills Trainer (EST), which are used to train small-unit leaders and crews. The constructive environment consists of battle staff training for battalion and above staff and leaders such as Joint Conflict and Tactical Simulation (JCATS)/OneSAF.

CONSTRUCTIVE ENVIRONMENT

1-12. The constructive training environment consists of the following, which all occur in the culminating training phase:

- WARSIM (division/corps), which includes warfighting exercise (battle command training program).
- JCATS/OneSAF (brigade and below), staff exercises (STAFFEXs), command post exercises (CPXs), and joint training exercises (JTX)

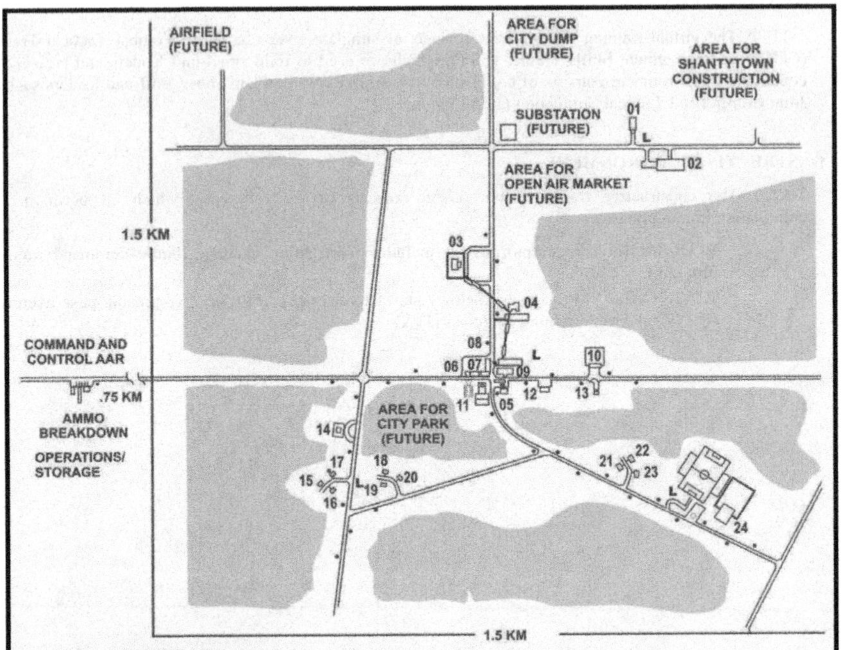

BLDG #	SIZE AND TYPE
01	1 STORY WAREHOUSE
02	1 STORY WAREHOUSE
03	3 STORY OFFICE (EMBASSY)
04	2 STORY OFFICE
05	1 STORY SERVICE STATION
06	1 STORY BUSINESS
07	1 STORY BUSINESS
08	1 STORY BUSINESS
09	3 STORY HOTEL
10	1 STORY POLICE STATION/JAIL
11	CHURCH AND CEMETARY
12	1 STORY BUSINESS
13	1 STORY BANK
14	2 STORY TOWNHOUSE
15	1 STORY RESIDENT
16	1 STORY RESIDENT
17	1 STORY RESIDENT
18	1 STORY RESIDENT
19	1 STORY RESIDENT
20	1 STORY RESIDENT
21	1 STORY RESIDENT
22	1 STORY RESIDENT
23	1 STORY RESIDENT
24	2 STORY SCHOOL

LEGEND

- FENCE WALL
- WALL
- TUNNEL SYSTEM
- TREE LINE
- UTILITY POLE
- LATRINE

SCALE: 1" = 500'

Figure 1-2. Combined Arms Collective Training Facility.

MISSION-ESSENTIAL TASK LIST

1-13. FMs 7-0 and 7-1 explain how to determine a unit's mission-essential task list (METL) and battle tasks, and how to plan, resource, and execute training events. Because training facilities are limited, leaders must use all available time and resources for training and maintaining combat readiness.

- System-specific FMs and TCs teach doctrine, conduct of operations, and employment of weapon systems.
- Tactics, techniques, and procedures (TTP) are lessons learned from training and combat.

1-14. The standards for evaluating training are accessible at the Digital Training Management System (DTMS), https://dtms.army.mil/DTMS.

BATTALION INTEGRATED TRAINING PLAN

1-15. The commander uses the battalion integrated training plan to set goals for training density. He focuses on combined arms operations, emphasizing dismounted Infantry maneuver. Tanks, BFVs, Strykers, attack aviation, and artillery normally provide supporting fires for the dismounted elements. UO training in the CACTF, Shoot House, and Breach Facility should be conducted semiannually and quarterly for the UAC. Army training calendars are divided into two cycles of training: *mission* and *support*. Figure 1-3 and Figure 1-4 show how the urban training strategy and facility usage might be programmed into unit cycles.

RISK ASSESSMENT

1-16. Risk assessment means making operations safe without compromising the level of realistic training (FM 7-1). Commanders continuously assess the risk of training conditions to prevent unnecessary loss of Soldiers and equipment. The degree of risk depends on conditions. Just as this principle applies to other training and warfighting, it also applies to UO. In a training situation, for example, commanders determine whether Soldiers and leaders have conducted this type of training before, whether conducting initial training at night or during limited visibility, and whether Soldiers are tired or exhausted. Risk assessment ensures UO training is not only tough and realistic but also safe. Training enhancers, such as live fire, blank ammunition, simulators, fires, smoke, and other pyrotechnics, must be fully integrated into the training to achieve the training objectives. Soldiers are required to wear earplugs, protective vests, and protective eyewear in order to protect themselves from noise and debris.

1-17. The commander considers risk and safety in urban operations training situations such as--.

- Soldiers engaging in close combat urban operations need to be safety conscious when using blanks, booby traps, and practice grenades. They may be unaware of the exact location of the OPFOR and may be susceptible to OPFOR flash suppressors, booby traps, and practice hand grenades.
- Soldiers should use ladders to enter buildings at upper stories of buildings to prevent unnecessary risks.

1-18. The commander is the safety officer and is responsible for ensuring that his unit, both leaders and Soldiers, comply with safety regulations and the units tactical SOP. All leaders must--.

- Use mission, enemy, terrain, troops, time available, and civilians (METT-TC) factors to identify risks.
- Assess possible losses and their costs.
- Select and develop risk-reduction measures.
- Implement controls by integrating them into plans and orders, SOPs, training performance standards, and rehearsals.
- Supervise and enforce risk reduction measures and safety standards at all times.

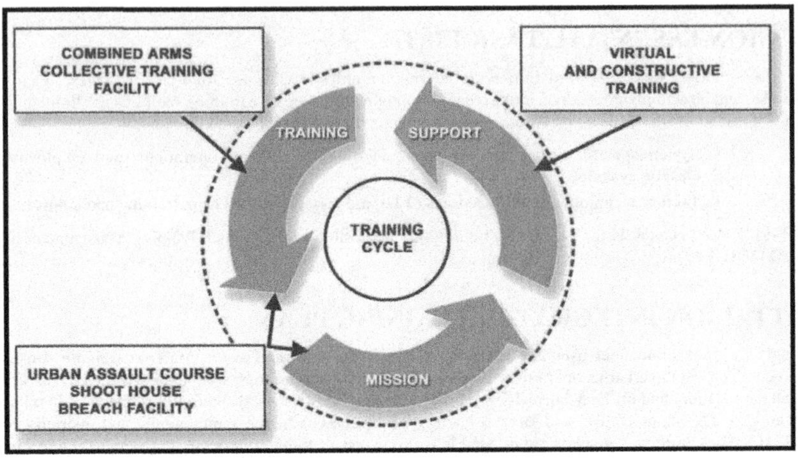

Figure 1-3. Active Army, Home Station, Urban Operations Training Strategy.

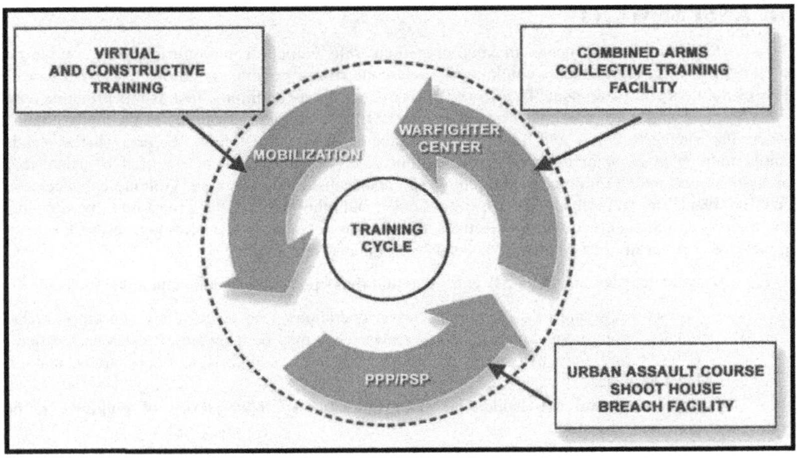

Figure 1-4. Army National Guard Urban Operations Training Strategy.

Section III. TRAINING FACILITIES

Training facilities include three that are discussed in this section: the Urban Assault Course (UAC, Chapter 2), Shoot House (SH, Chapter 3), Breach Facility (BF, Chapter 4). They also include the Combined Arms Collective Training Facility (CACTF, discussed in Chapter 5), Combat Training Centers (CTC), and any other live training that supports the UO training.

URBAN ASSAULT COURSE

1-19. The purpose of the UAC facility is to train a squad-sized force to *Enter a Building and Clear a Room*, improve grenadier gunnery, and validate subterranean TTP. However, the unit's training requirements determine the size of the unit as well as the tasks to be trained. The unit develops its organizational structure and steps for conducting the exercise based on its METL and on their training needs (Figure 1-5). (See Chapter 2 for a more detailed discussion.)

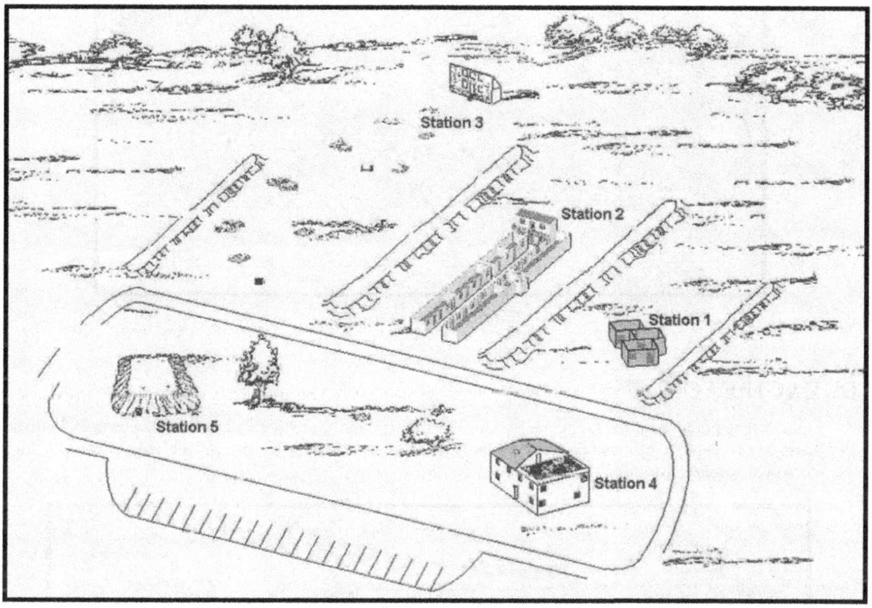

Figure 1-5. Urban Assault Course.

SHOOT HOUSE

1-20. The purpose of this facility is to provide the leader with a facility to train and evaluate fire teams, squads, and platoons during a live-fire exercise (Figure 1-6). Units are trained and evaluated on their ability to conduct tactical movement and engage targets, while practicing target discrimination in an urban environment. (See Chapter 3 for a more detailed discussion.)

Note: The specific designs of the Shoot House and the other facilities may differ from one installation to another, due to variations in training requirements.

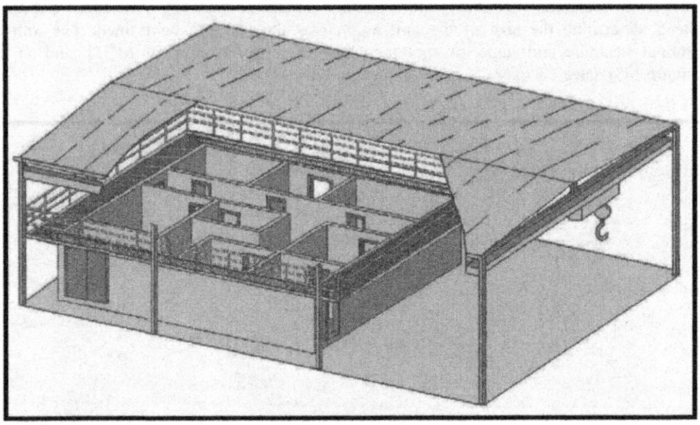

Figure 1-6. Shoot House.

BREACH FACILITY

1-21. The purpose of this facility is to train Warfighters on the technical aspects of mechanical, ballistic, thermal, and explosive breaching techniques (Figure 1-7), including for use in UO. (See Chapter 4 for a more detailed discussion.)

Figure 1-7. Breach Facility.

Section IV. CONDUCT OF TRAINING

Depending on unit proficiency, every training event should follow a *crawl-walk-run* sequence, based on clearly defined task, conditions, and standards. After each training event, leaders conduct an after-action review (AAR).

CRAWL

1-22. The trainer explains each training objective and performance standards, and then talks the Soldiers through the exercise, step by step, describing what each individual or unit must do. This training often includes a demonstration.

WALK

1-23. The Soldiers slowly practice each task to standard. Trainers coach Soldiers at this stage stopping as often as needed to correct mistakes and provide feedback. Soldiers practice each task repeatedly until they can perform it to standard.

RUN

1-24. Soldiers perform each task at full speed as if they are in combat. The Soldiers need feedback during this stage as well. The following can be used to enhance realism:

- Blank ammunition.
- Special Effects Small-Arms Marking System (SESAMS).
- Targetry.
- Pyrotechnics.
- Tactical engagement systems (MILES).
- Opposing forces (force on force).
- Nuclear, biological, and chemical simulation/weapons of mass destruction.
- Training in limited visibility conditions.

AFTER-ACTION REVIEW

1-25. Leaders conduct an after action review (AAR) after each training event to help Soldiers and units improve their skills by providing immediate feedback. An AAR increases the benefits gained from each training exercise by allowing the leaders and the Soldiers to work together to analyze the performance of each task. Being involved in this analysis enhances the leaders and the Soldiers learning and retention ability. The AAR improves the performance of the leaders and Soldiers in later iterations by allowing interaction and feedback from these individuals during the debrief.

Section V. PREREQUISITE TRAINING

Prerequisite training teaches individuals, crews, units, leaders, and command and staff what they need to know before they are required to perform a task. Prerequisite training builds teamwork and individual confidence at all levels.

COMMAND AND STAFF TRAINING

1-26. The battalion battle staff and leadership must train before the battalion is required to conduct collective training. The battalion command and staff culminates their prerequisite training with a tactical exercise without troops (TEWTs) and command post exercises (CPXs). This sequence of training assists the commander and his staff develop better UO TTP and unit SOP.

LEADER TRAINING

1-27. A comprehensive leader's training program is the key to successful UO training (Appendix D). Time spent teaching and training the battalion leaders (trainers) during the planning and preparation phase will ensure success during the intermediate and culminating phases of training. Realistic training at the unit level will improve the leadership training officers and NCOs receive at the Basic Officer Leadership Course (BOLC), Captains Career Officer Course (CCOC), Basic Noncommissioned Officer Course [BNCOC], and Advanced Noncommissioned Officer Course [ANCOC]. Recommended training should include remedial training on those tasks that the unit had difficulty in during the previous training and evaluation.

INDIVIDUAL TRAINING

1-28. Many individual specific tasks and some collective tasks can be taught and practiced in garrison while most collective tasks are best conducted at the installation UO facilities and weapons ranges. The unit's preliminary individual training tasks should focus on three general areas: urban combat skills, urban marksmanship skills, and physical fitness. The following are example preliminary training tasks or TTP:

- Individual movement techniques.
- Battle drills.
- Weapon positioning in an urban area.
- Roadblock and vehicle search procedures.
- Dismounted and mounted urban navigation.
- Urban scanning, quick-fire, and assault fire techniques.
- Hasty urban firing positions.
- Preparation of urban firing positions.

1-29. Prior to urban collective training cycle, trainers should develop urban-specific firing conditions in conjunction with weapons firing ranges. Build facades on existing ranges, or use the unit's local UAC (if available). Urban combat skills include--

- Firing positions behind rubble, around corners, from windows or rooftops.
- Urban targets behind rubble, in windows and doorways.
- Fire commands.
- Lifting and shifting fires from lower to upper story windows.
- Fire control and fire discipline.
- Rapid firing engagements.
- ROE firing scenarios.

1-30. Urban physical fitness skills include--

- Endurance runs and upper body conditioning.
- Vaulting exercises.

- Urban-specific obstacle course negotiation.
- Leadership Reaction Course negotiation.

INDIVIDUAL TASK TRAINING

1-31. Urban operations require the Soldier to be proficient in those individual tasks that are unique to urban combat. Units should incorporate into their training regime those urban fighting TTP that have proven successful based on lessons learned from combat operations.

TASK FORCE SOLDIER--WARRIOR COMMON TASKS, SKILL LEVELS 1-3:

- 071-326-0541, *Perform Movement Techniques during an Urban Operation.*
- 071-326-0542, *Enter a Building during an Urban Operation.*
- 071-326-0550, *Prepare Positions for Individual and Crew-Served Weapons during an Urban Operation.*
- 071-326-0557, *Select Hasty Firing Positions during an Urban Operation.*
- 071-440-0027, *Execute Mechanical Breach in an Urban Operation.*
- 071-440-0028, *Engage Targets during an Urban Operation.*
- 071-440-0029, *Conduct a Breach during an Urban Operation.*
- 071-440-0003, *Conduct an Attack on a Building by a Squad during an Urban Operation.*
- 071-440-0006, *Conduct a Defense by Squad during an Urban Operation.*

FM 3-06.11, *COMBINED ARMS OPERATIONS IN URBAN TERRAIN*

Movement Techniques

- Crossing of a wall.
- Movement around corners.
- Movement past windows.
- Use of doorways.
- Movement parallel to buildings.
- Crossing open areas.
- Fire team employment.
- Movement between positions.
- Movement inside a building.

Entry Techniques

- Upper building levels.
- Use of ladders.
- Scaling of walls.
- Entry at lower levels.
- Hand grenades.

Firing Positions

- Hasty firing position.
- Prepared firing position.
- Target acquisition.
- Employment of snipers.

COLLECTIVE TASK TRAINING

1-32. Specific FMs teach doctrine, how to conduct operations, and how to employ weapon systems. TTP are lessons learned from the training and combat. The standards for conducting and evaluating collective training are accessible at the Digital Training Management System (DTMS), https://dtms.army.mil/DTMS. Those tasks that do not specifically address UO must be modified for use in an urban environment.

RESOURCES

1-33. Field manuals for the squad and platoon are--:

- FM 3-21.8, *Infantry Rifle Platoon and Squad.*
- FM 3-21.71, *Mechanized Infantry Platoon and Squad (BFV).*
- FM 3-21.9, *SBCT Infantry Rifle Platoon and Squad.*

1-34. Field manuals for the company are--.

- FM 3-21.10, *IBCT Rifle Company.*
- FM 3-21.11, *SBCT Rifle Company.*

1-35. Field manuals and mission training plans for the battalion are--.

- FM 3-21.20, *IBCT Infantry Battalion.*
- FM 3-21.21, *SBCT Infantry Battalion.*
- FM 3-06.11, *Combined Arms Operations in Urban Terrain.*
- TC 7-98-1, *Brigade and Battalion Stability and Support Operations.*

Chapter 2

Urban Assault Course

This chapter describes the Urban Assault Course (UAC), which incorporates doctrine and tactics from FM 3-06.11. The UAC contains five stations:

- Individual and Team Task/Technique Trainer Station 1
- Squad and Platoon Task/Technique Trainer Station 2
- Grenadier Gunnery Trainer Station 3
- Urban Offense/Defense Building Station 4
- Underground Trainer Station 5

These stations are designed for individual tasks and small-unit collective task training. Various types of organizations with a variety of missions can train at the UAC. Tactical operations in an urban environment may include combat to stability and support tasks. (Appendix C provides a SASO unit training matrix). The UAC is a dismounted training facility.

Note: The station discussions are arranged so that they may be removed from the book for ease of use.

PURPOSE

2-1. The purpose is to provide squad and platoon size units with a facility to train and evaluate urban operations tasks (Figure 2-1). Only Station 3, Grenadier Gunnery, is intended for live-fire training. It supports use of 40-mm TP and 5.56-mm service ammunition.

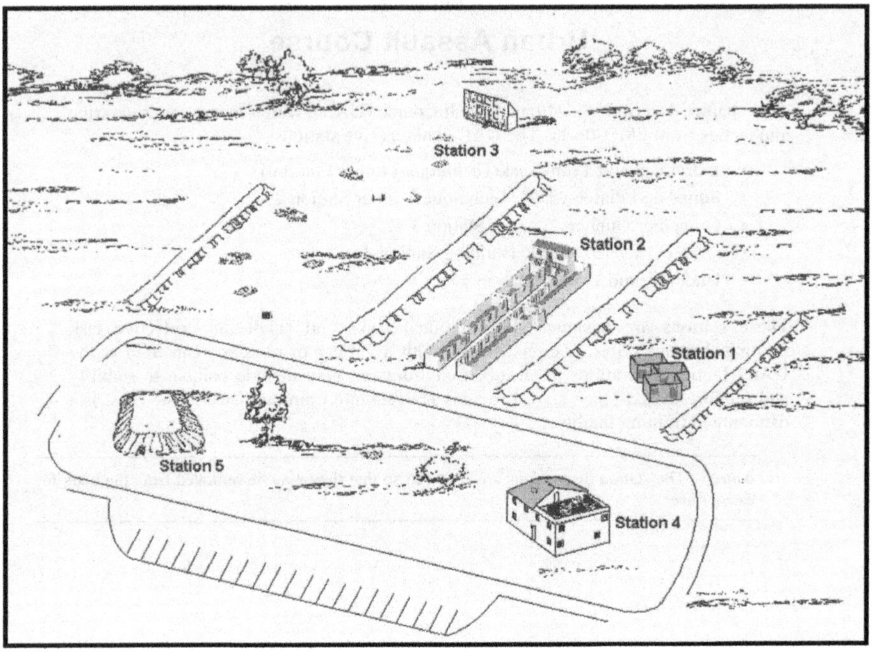

Figure 2-1. Urban Assault Course.

DESCRIPTION

2-2. The UAC has five training stations that facilitate the ***crawl-walk-run*** training concept:

Station 1, Individual and Team Trainer

2-3. This station has three rooms where team leaders and squad leaders train the basics of building and room clearing.

Station 2, Squad and Platoon Trainer

2-4. This station has four structures and multiple rooms. Squads build on tasks learned at Station 1, and begin to learn the concepts of clearing multiple buildings. This station can be used as individual buildings with a narrow street, or as rooms inside a building with a long, connecting hallway.

Station 3, Grenadier Gunnery Trainer

2-5. This is a live-fire station where M203 gunners master target engagements in an urban area, conduct tactical movement, and respond to the fire commands.

Station 4, Offense/Defense House

2-6. This is where a platoon trains to attack or defend a building. This station can also be divided into smaller training stations to reinforce training. It is also used to train tasks not yet trained at the other stations such as upper-level entry techniques.

Station 5, Underground Trainer

2-7. This station provides training for subterranean operations.

TARGETRY

2-8. Targets throughout the UAC are either 3D precision targets (for engagements less than 50 meters) or 2D nonprecision targets (for engagements greater than 50 meters).

Precision Targets

2-9. Precision targets support short distance engagement sensors. These sensors, located inside the precision targets, respond only to lethal shot placements.

- Torso lethal zone .... 18" wide x 8" high
- Neck-head lethal zone .... 4" wide x 8" high

Nonprecision Targets

2-10. Nonprecision targets support engagements at ranges greater than 50 meters. These targets are 2D pop-up type targets.

Training Audience

2-11. The facility gives Infantry, Engineer, Military Police, Cavalry, and other units a facility for training individual Soldiers through platoon-level UO skills.

Training Information

2-12. Combat units can use this facility as a mission rehearsal exercise site. Recommended training frequency is quarterly. For each UAC station, they should provide the purpose, a description, targetry, controller and safety requirements, individual UO tasks to be trained, training support requirements, and conduct of training information, if applicable:

Purpose

2-13. This UAC station [*state purpose of station*]. [*Units may use the station in the manner originally intended, or they may modify it to meet their unique training needs.*]

Description

2-14. This station is [*describe station use and design*].

Targetry

2-15. Recommended targetry for training at this station includes [*describe*]. Targets may be moved, modified, or added, depending on available target systems or training requirements.

Controller

2-16. Controller requirements for this station include [*describe*].

Safety

2-17. Safety requirements and considerations for this station include [*describe*]. [*The unit or installation range SOP may add more stringent safety requirements.*]

Individual Tasks or Skills Trained

2-18. These may be found in the Digital Training Management System located online at https://dtms.army.mil/DTMS.

Training Support

2-19. Minimum materiel requirements to support training at this station include [describe].

Conduct of Training

2-20. Example training scenarios and sequence of events for this station include [describe]. [The UAC can accommodate an assortment of training ammunition and pyrotechnics.]

STATION 1: INDIVIDUAL AND TEAM TASK/TECHNIQUE TRAINER

2-21. This station is used to train individual and collective tasks, tactics, techniques, and procedures, specifically *Enter a Building and Clear a Room* and *Engage Targets* (Figure 2-2). It has three rooms where team leaders and squad leaders train the basics of building and room clearing.

DESCRIPTION

2-22. Station 1 has a single wood structure. Its three rooms have one or more open doorways and windows, and a man-sized hole.

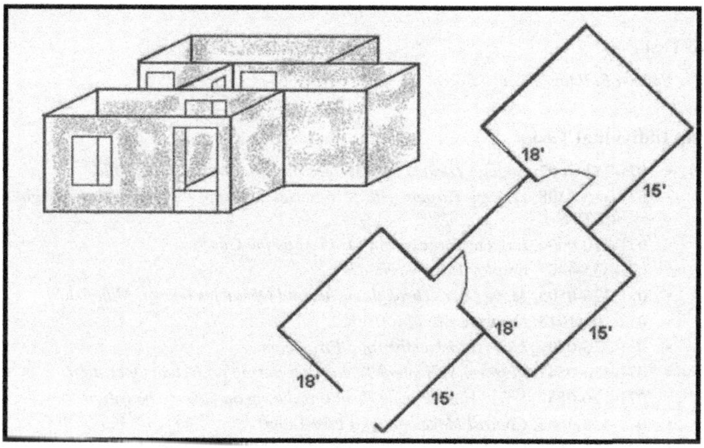

Figure 2-2. Individual and team task/technique trainer, Station 1.

TARGETRY

2-23. Targets present realistic close quarters combat situations (0 to 15 meters), which consist of a possible mixture of combatant and noncombatant targets. Target discrimination is trained by presenting up to six precision targets. Targetry sensors respond only to lethal shot placement. They also aid in training during limited visibility.

CONTROLLER REQUIREMENTS

2-24. These are based on unit requirements.

SAFETY REQUIREMENTS

2-25. Only smoke and practice grenades, or M84 stun grenades, may be used at this station.

> **WARNING**
>
> **To avoid injury, place the weapon on SAFE when climbing through obstacles such as windows, fences, and walls during training.**

INDIVIDUAL/COLLECTIVE TASKS TRAINED

2-26. This is noninclusive.

Collective Task

2-27. *Enter a Building/Clear a Room.*

Supporting Individual Tasks

- 071-311-2007, *Engage Targets with M16-Series Rifle/M4 Carbine Rifle.*
- 071-315-2308, *Engage Targets with M16-Series Rifle/M4 Carbine Using a Night Vision Sight AN/PVS-4.*
- 071-010-0006, *Engage Targets with M249 Machine Gun.*
- 071-325-4407, *Employ Hand Grenades.*
- 071-326-0503, *Move Over, Through, or Around Obstacles (except Minefields).*
- 052-193-1013, *Neutralize Booby Traps.*
- 071-326-0501, *Move as a Member of a Fire Team.*
- 071-326-0541, *Perform Movement Techniques during an Urban Operation.*
- 071-326-0557, *Select Hasty Firing Positions during an Urban Operation.*
- 071-326-5605, *Control Movement of a Fire Team.*
- 071-326-5611, *Conduct the Maneuver of a Squad.*
- 071-710-0008, *Operate Night Vision Goggles AN/PVS-7B.*

Related Tactics, Techniques, and Procedures

- FM 3-06.11, *Combined Arms Operations in Urban Terrain.*
- FM 3-22.9, *Rifle Marksmanship.*

TRAINING SUPPORT REQUIREMENTS

2-28. Use Table 2-1 for the minimum ammunition required to train at this station.

Table 2-1. Support requirements for Station 1.

Ammunition	*Rounds per Weapon*
5.56-mm Blank	10 rounds
5.56 mm Blank (Linked)	15 rounds
Practice Grenade Bodies	2 per team
Practice Grenade Fuses	2 per team

CONDUCT OF TRAINING

2-29. Table 2-2 lists the training sequence and times for Station 1.

Table 2-2. Sequence of events and times for Station 1.

Event	*Action*	*Estimated Time*
1	*Employ Hand Grenades*	30 minutes crawl, walk, run
	Perform Movement Techniques • *Engage Target with M16* • *Target Discrimination*	20 minutes crawl 20 minutes walk 10 minutes run
2	*Enter a Building*	20 minutes crawl 20 minutes walk 10 minutes run
3	*Enter and Clear a Room*	20 minutes crawl 20 minutes walk 10 minutes run
4	*Conduct an AAR*	40 minutes

Example Scenarios, Assault Team, Varied Entry Positions

2-30. Figures 2-3 and 2-4 show example scenarios. Rather than using the examples, units should tailor their training events based on their unit's METL and training needs.

Varied Entry Positions

2-31. The team leader first trains to execute *Varied Entry Positions* without targets, which is part of the *crawl* phase of training. The trainer has the assault teams position at various places to demonstrate how the start position can affect the final points of domination. (Figure 2-3).

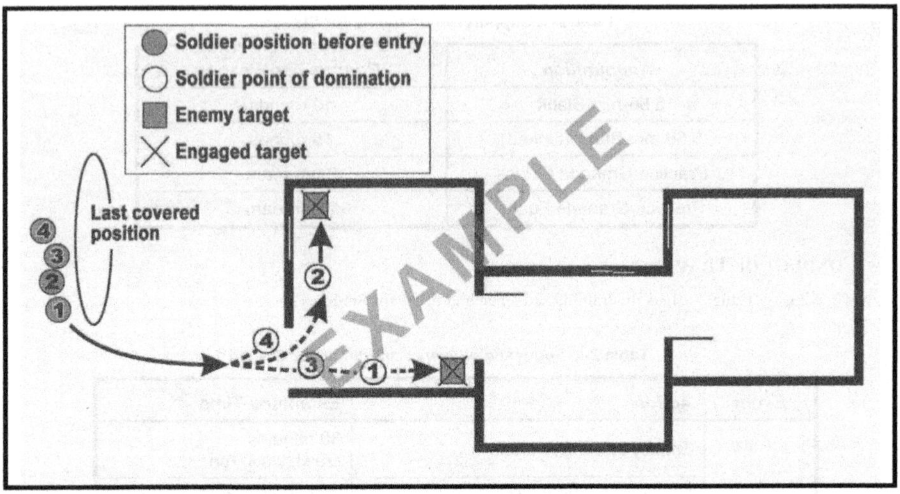

Figure 2-3. Example Station 1 scenario, *Varied Entry Positions.*

Multiple Room-Clearing Techniques

2-32. The team leader trains his Soldiers to execute *Varied Entry Positions* using targets in a multiple room scenario. This is part of the *walk* phase. The trainer has the assault teams enter the room and engage targets. He has them use using precision marksmanship techniques with the Tactical Engagement System (TES) or SESAMS (Figure 2-4). The assault team enters the first room and engages the targets. As the number two Soldier clears his area of responsibility, he sees and engages the target in the second room, but remains in place until the assault teams finishes clearing the first room. Once the room is clear, the assault team moves to the second room and continues clearing, or lets the next team move through.

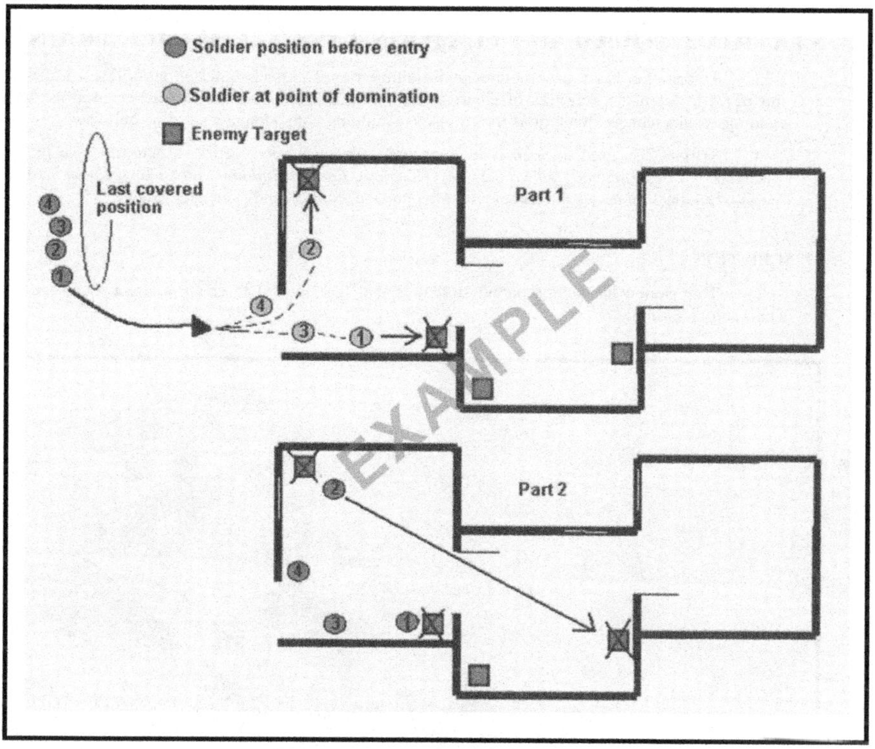

Figure 2-4. Example Station 1 scenario,
Multiple Room-Clearing Techniques.

STATION 2: SQUAD AND PLATOON TASK/TECHNIQUE TRAINER

2-33. This station has four structures and multiple rooms. Squads build on tasks learned at Station 1, and begin to learn the concepts of clearing multiple buildings. This station can be used as individual buildings with a narrow street, or as rooms inside a building with a long, connecting hallway.

2-34. Station 2 is used to train individual and collective tasks, tactics, techniques, and procedures, specifically, *Move Tactically Down a Street (Hallway)*, *Enter a Building and Clear a Room*, and *Engage Targets*. Station 2 increases the complexity of command, control, and maneuver (Figure 2-5).

DESCRIPTION

2-35. This station has four separate structures that face inward to create a street. It also introduces a two-story building.

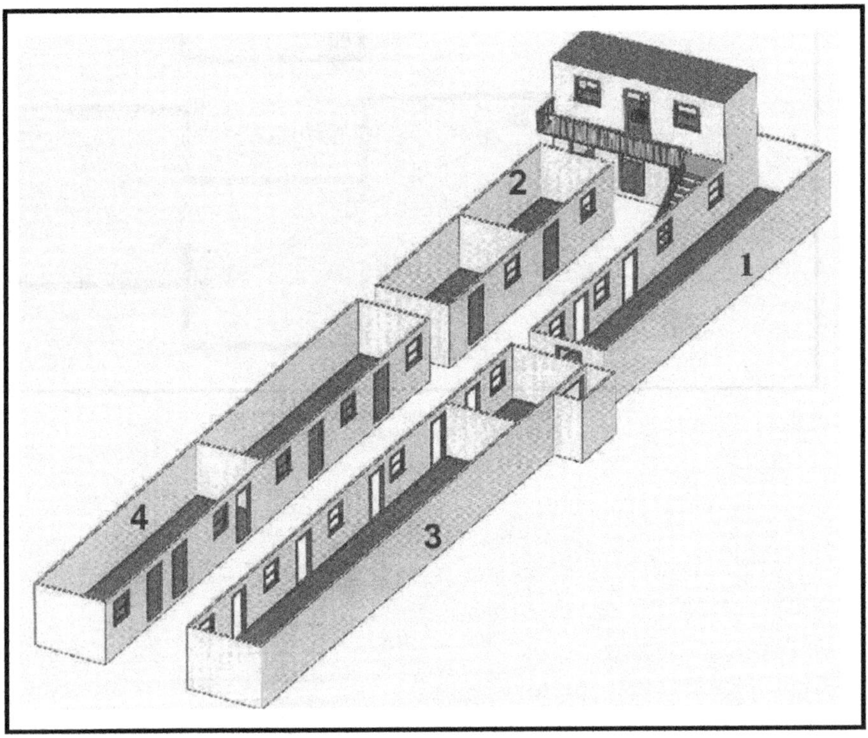

Figure 2-5. Squad and platoon task and technique trainer, Station 2.

> Building 1 is L-shaped.
>
> - The long leg of the building to the right looks like a single-story building.
> - The short leg of the building to the front has two floors. A landing along the length of the second floor has outside stairs. An inside stairway also joins the floors.
> - The two legs of the "L" are adjacent, but without interior access to each other.
>
> Building 2 has a single floor and two rooms, each with a door and window. An open area much like a street separates this structure from the first.
>
> Building 3 is a row of wooden buildings along a street. It has many windows and doorways. At the far end is an adjoining room.
>
> Building 4 is on the left. It mirrors Building 3, plus it has a wall on the side at the near end.

2-36. Units can place additional walls in any structure by using target cloth or other material. They can do this to add to the number of rooms or to increase the complexity of the training (Appendix A).

TARGETRY

2-37. Station 2 targets present realistic close quarters combat situations (0 to 15 meters), with the option to customize the mixture of combatant and noncombatant targets. Target discrimination is trained on up to ten precision targets. Targetry sensors respond only to lethal shot placement, and facilitate training during limited visibility.

CONTROLLER REQUIREMENTS

2-38. These are based on unit requirements.

SAFETY REQUIREMENTS

2-39. At this station, Soldiers may only use smoke grenades, practice grenades, and M84 stun grenades.

> **WARNING**
>
> **To avoid injury, place the weapon on SAFE when climbing through obstacles such as windows, fences, and walls during training.**

INDIVIDUAL AND COLLECTIVE TASKS TRAINED

2-40. This list is noninclusive.

Collective Task--*Enter and Clear a Building*

Related Collective Tasks

- *Conduct Consolidation and Reorganization (Company/Platoon).*
- *Treat Casualties.*
- *Evacuate Casualties.*
- *Defend MOUT Building.*

- *Conduct Tactical Movement.*
- *Enter and Clear a Room.*

Supporting Individual Tasks

- 052-193-1013, *Neutralize Booby Traps.*
- 071-010-0006, *Engage Targets with M249 Machine Gun.*
- 071-311-2007, *Engage Targets with M16-Series Rifle/M4 Carbine Rifle.*
- 071-315-2308, *Engage Targets with M16-Series Rifle/M4 Carbine Using a Night Vision Sight AN/PVS-4.*
- 071-325-4407, *Employ Hand Grenades.*
- 071-326-0501, *Move as a Member of a Fire Team.*
- 071-326-0503, *Move Over, Through, or Around Obstacles (except Minefields).*
- 071-326-0541, *Perform Movement Techniques during an Urban Operation.*
- 071-326-0557, *Select Hasty Firing Positions during an Urban Operation.*
- 071-326-0608, *Use Visual Signaling Techniques.*
- 071-326-5605, *Control Movement of a Fire Team.*
- 071-326-5611, *Conduct the Maneuver of a Squad.*
- 071-710-0008, *Operate Night Vision Goggles AN/PVS-7B.*
- 113-571-1022, *Perform Voice Communications.*
- 191-377-5250, *Handle Enemy Personnel and Equipment.*

Related Tactics, Techniques, and Procedures

- FM 3-06.11, *Combined Arms Operations in Urban Terrain.*
  -- Chapter 3, *Offensive Operation.*
  -- Chapter 5, *Fundamental Combat Skills.*
  -- Appendix M, *Field-Expedient Breaching of Common Urban Barriers.*
- FM 3-22.9, *Rifle Marksmanship.*

TRAINING SUPPORT REQUIREMENTS

2-41. Table 2-3 shows the minimum ammunition required to train at this station.

Table 2-3. Support requirements for Station 2.

Ammunition	Rounds per Weapon
5.56-mm blank	16
5.56-mm blank (linked)	24
Practice grenade fuses	8 per squad
Practice grenade bodies	8

Conduct of Training

2-42. Table 2-4 shows the sequence of events and times for Station 2. Figure 2-6 through Figure 2-28 show example scenarios. However, rather than use these examples, the unit should customize its training events based on its METL and training needs.

Table 2-4. Sequence of events and times for Station 2.

Event	Action	Estimated Time
1	*Conduct Tactical Movement in a Built-Up Area* *Move Across an Open Area* *Move Parallel to Buildings*	10 min Crawl 20 min Walk 10 min Run
2	*Enter and Clear a Room (Squad)* *Clear a Stairwell* *Clear a Hallway* *Conduct a Breach*	10 min Crawl 20 min Walk 10 min Run
3	*Enter and Clear a Building (Platoon)* *Clear Multiple Rooms* *Consolidate and Reorganize*	10 min Crawl 20 min Walk 10 min Run
4	*Conduct AAR (Squad)*	10 min
NOTE: Training should include techniques outlined FM 3-06.11.		

2-43. The squad is clearing Station 2 as if it were a single building, and the street a hallway. Fire Team A moves down the left side, throws the appropriate grenade IAW ROE, enters, engages the target, and clears the room on that side. The last Soldier remains outside and secures the rear Fire Team B prepares to enter the next room (Figure 2-6).

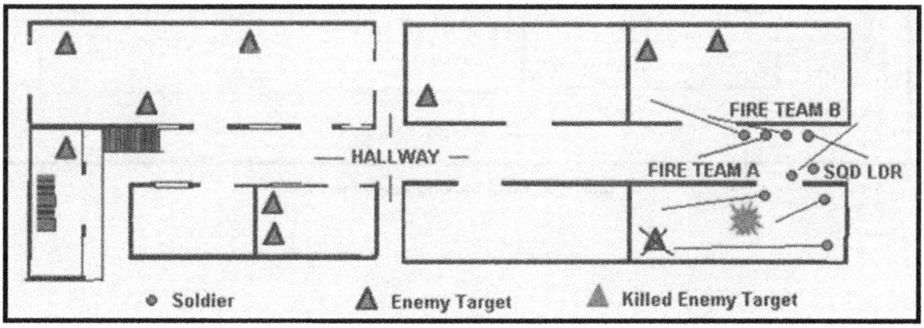

Figure 2-6. Conducting tactical movement.

2-44. Fire Team B throws the appropriate grenade, enters, engages the targets, and clears the room on the right while the last Soldier secures the rear Fire Team A prepares to continue movement to the next room (Figure 2-7).

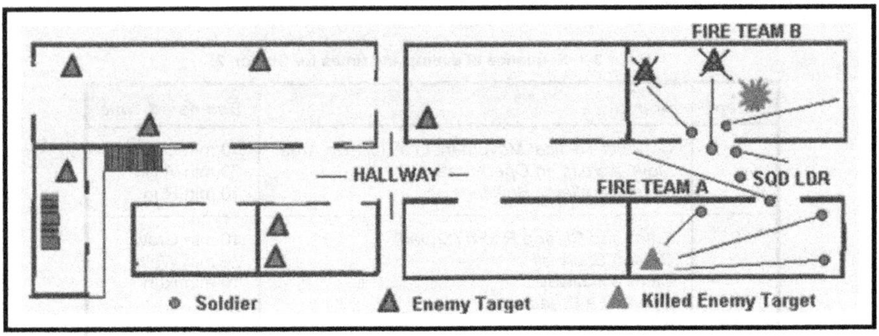

Figure 2-7. Team B--Clearing a building.

2-45. Fire Team A moves across the hall, throws the appropriate grenade, enters, engages the target, and clears the next room Three team members enter the room, the fourth remains in the doorway and secures the hallway (Figure 2-8).

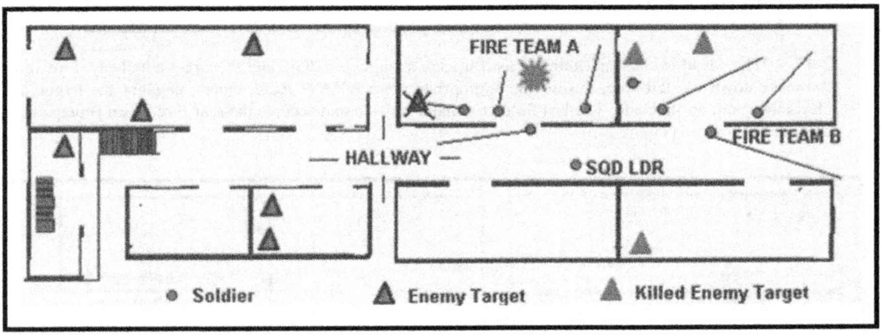

Figure 2-8. Team moving across the hall.

2-46. Fire Team B moves in a cross-cover formation down the hallway to the next room, throws the appropriate grenade, and clears the room. As Fire Team B passes the security man of Team A, he turns and becomes rear security. The last Soldier in Fire Team B remains by the door and secures the hallway (Figure 2-9).

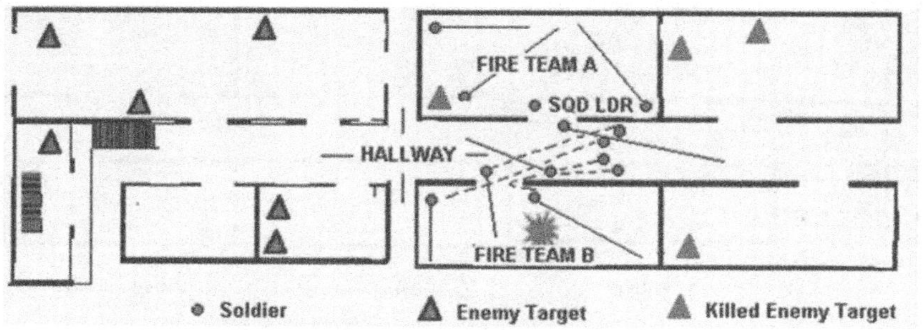

Figure 2-9. Team B--Moving in cross cover formation.

2-47. Fire Team A clears the hallway intersection, while Team B uses the Rolling-T technique to pass through Team A en route to the next room (Figure 2-10).

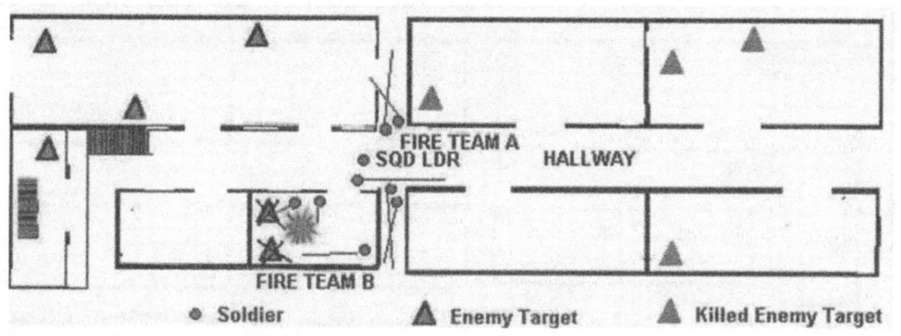

Figure 2-10. Team A--Clearing a hallway intersection.

2-48. Fire Team B throws the appropriate grenade, enters, engages the targets, and clears the room. Only three Soldiers enter, while the fourth Soldier provides rear security. Fire Team A prepares to move (Figure 2-11).

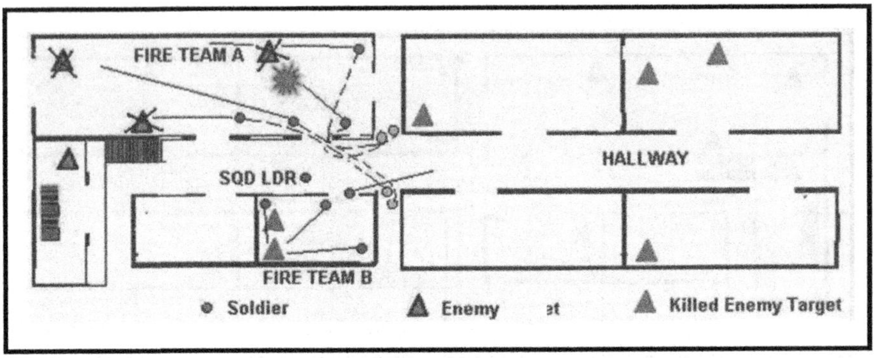

Figure 2-11. Team B--Clearing a room.

2-49. Fire Team A moves from the intersection and throws the appropriate grenade. The team enters, engages the targets, and then clears the next room. All team members enter, because of the size of the room. Fire Team B prepares to move (Figure 2-12).

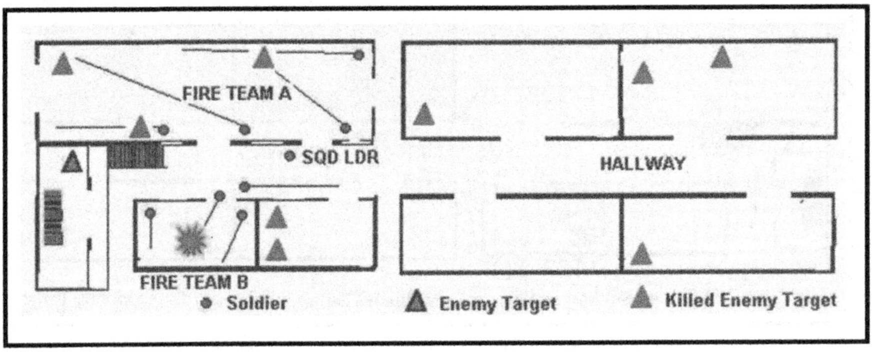

Figure 2-12. Team A--Clearing room by room.

2-50. Fire Team B employs the appropriate grenade, enters, and engages the target. Three Soldiers clear the next room. The fourth remains in the doorway and secures the rear. Fire Team A prepares to move up the stairs (Figure 2-13).

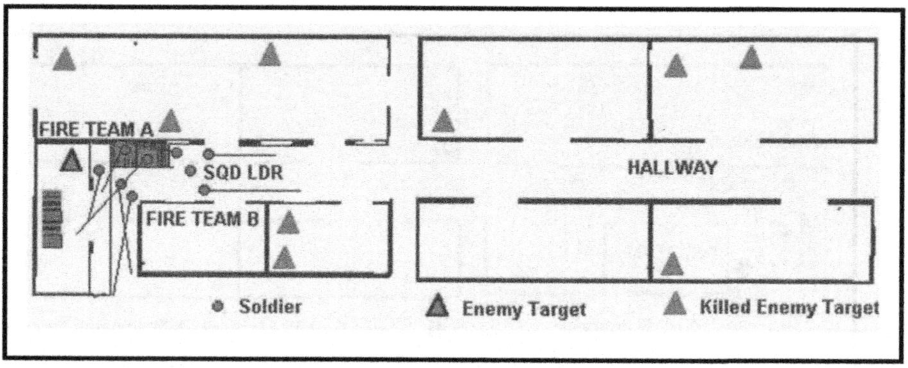

Figure 2-13. Team B--Clearing the next room.

2-51. Fire Team A moves up the stairs and onto the landing. The team prepares to enter the room off the landing. Fire Team B splits, and two members move around the corner and secure the area across from the stairs on the lower level. The other two members of Team B secure the hallway the squad came from (Figure 2-14).

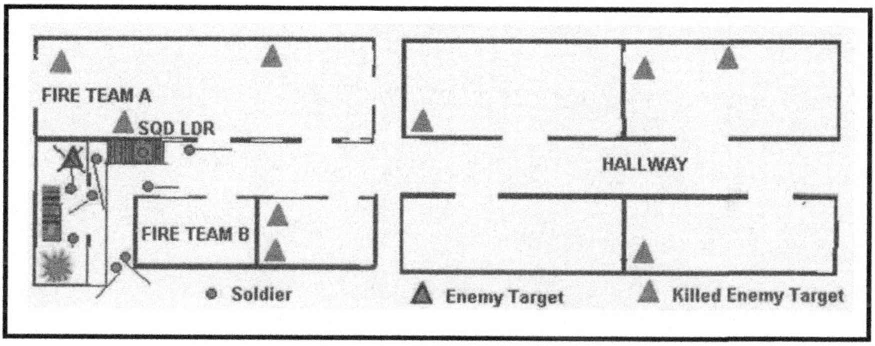

Figure 2-14. Securing the stairs and hallway.

2-52. Fire Team A members employ the appropriate grenade, enters the room, engages the target, and clears it. They find another set of stairs, and prepare to move up them. Fire Team B continues to secure the rear. The other team members move to the end of the short hallway and secure both directions (Figure 2-15).

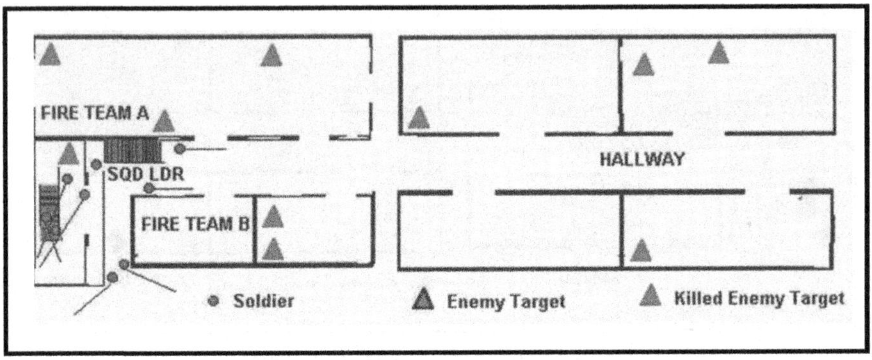

Figure 2-15. Preparing to clear more stairs.

2-53. Fire Team A clears the stairs and the landing. Fire team B continues to secure the ground level, ready to assist Team A, if necessary (Figure 2-16).

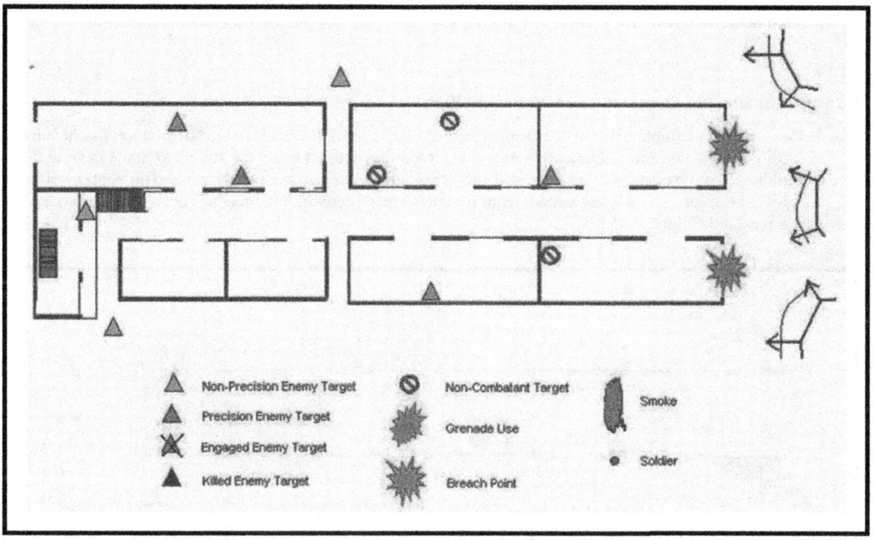

Figure 2-16. Team A--Clearing stairs and landing.

2-54. The previous example used no noncombatant targets, but used blanks/SESAMS and TES. This technique is the *crawl* phase for this station. Once the squad is ready to move on to more challenging exercises, the platoon leader/sergeant can add targets into the scenario. As proficiency increases, mixed combatant and noncombatant targets can be added. At any point, the platoon leader/sergeant can, until the platoon achieves proficiency, include multiple squads on the station. This increases the complexity of the scenarios.

Example Scenario--Preparing to Gain a Foothold

2-55. In this example platoon scenario, noncombatants are on the battlefield. The platoon has decided to clear through the buildings. They are going to try to do this without using the streets to move from building to building. Two squads will assault, and the other will remain in support/reserve. The support element isolates the buildings, while the assault team for each squad prepares and breaches each building in order to gain a foothold (Figure 2-17).

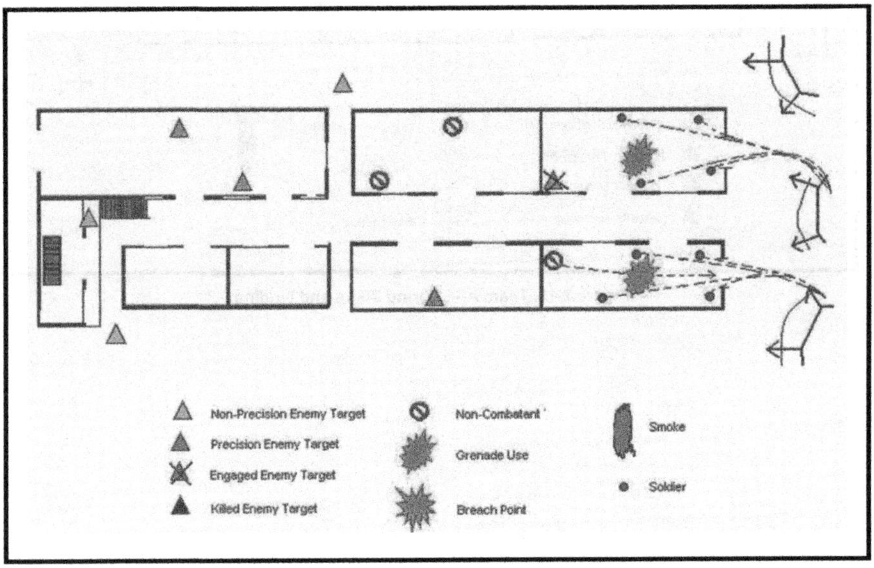

Figure 2-17. Preparing to gain a foothold.

2-56. The assault teams throw a grenade into the breach and their members follow, engaging enemy targets. They secure and move noncombatants to the collection point (CP) (Figure 2-18).

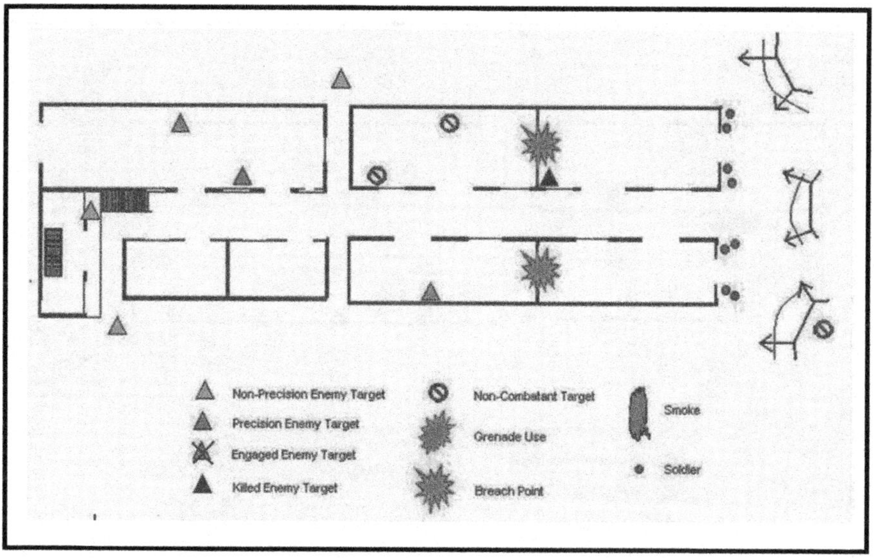

Figure 2-18. Securing and moving noncombatants to the command post.

2-57. The teams prepare and detonate wall breaches to continue movement through the building. Once the charges are ready for detonation, the assault teams move to a covered position and execute the breach (Figure 2-19).

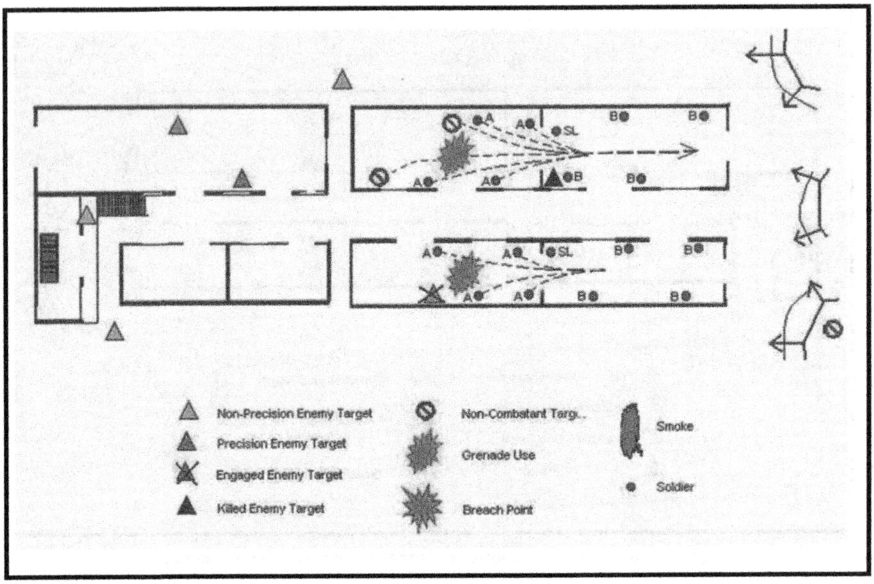

Figure 2-19. Executing breach.

2-58. The assault team employs grenades into the next room, enters the room, eliminates the enemy, and then secures and moves the noncombatants to a designated safe area. At the same time, the next assault team moves into the foothold (Figure 2-20).

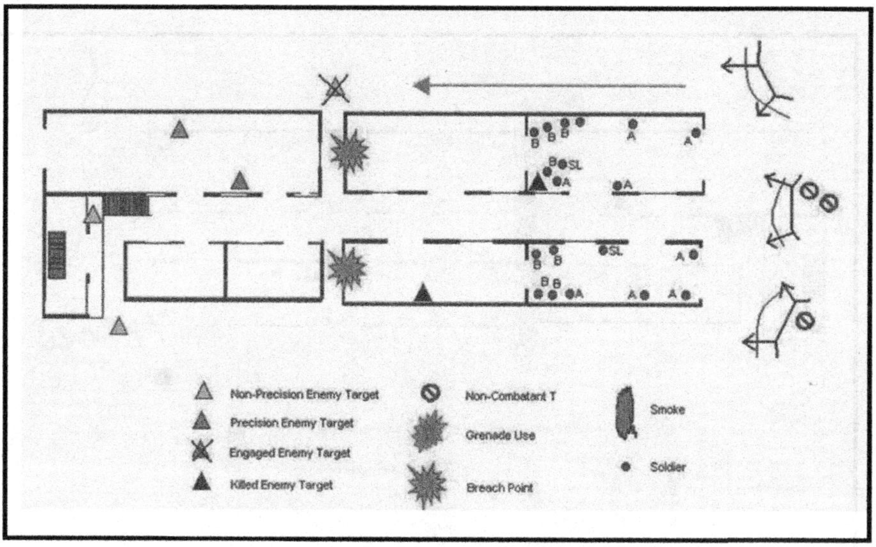

Figure 2-20. Moving to the foothold.

2-59. With the second room secured, another assault team moves forward and prepares to breach the wall Once the charges are placed, the teams move back to a covered position before detonating the charge. With the breach a success, the first nonprecision target is exposed and engaged by the support element (Figure 2-21).

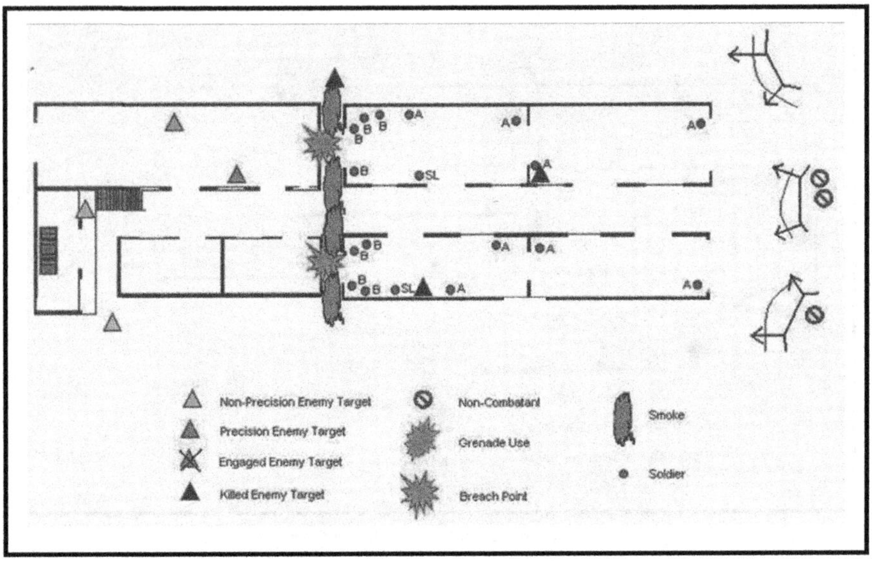

Figure 2-21. Breaching the wall.

2-60. Upon inspecting the breach, the assault teams find that they have an alley with another wall to breach. They employ smoke to obscure the teams placing the breaching charge. The first assault teams take up securing positions throughout the already cleared areas. Follow-on assault teams move to a covered position and execute the breach (Figure 2-22).

Note: Use smoke only in the alley, not inside the buildings.

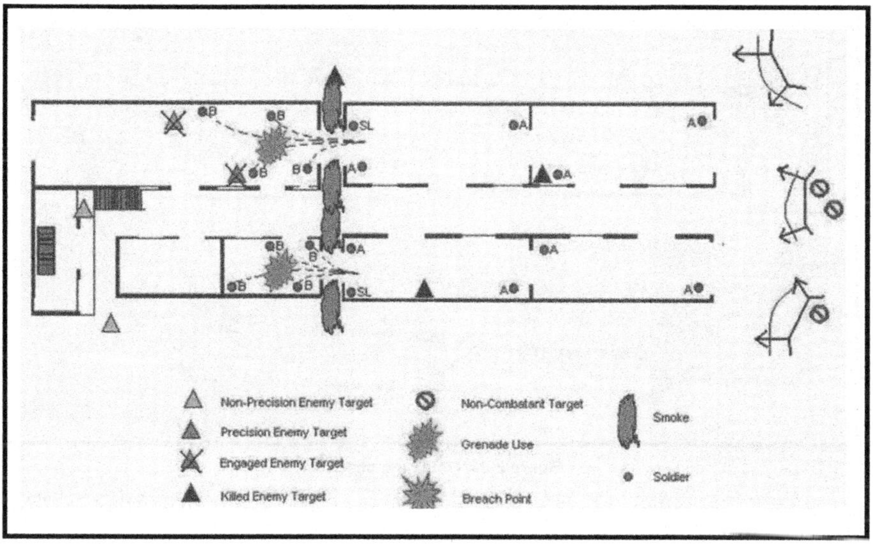

Figure 2-22. Securing a wall in an alley.

2-61. Once the assault teams execute the breach, they must determine if they need to use more smoke to cover their movement. Either way, once they have enough smoke to cover their movement, they throw in a grenade, enter, engage enemy targets, and clear the rooms (Figure 2-23).

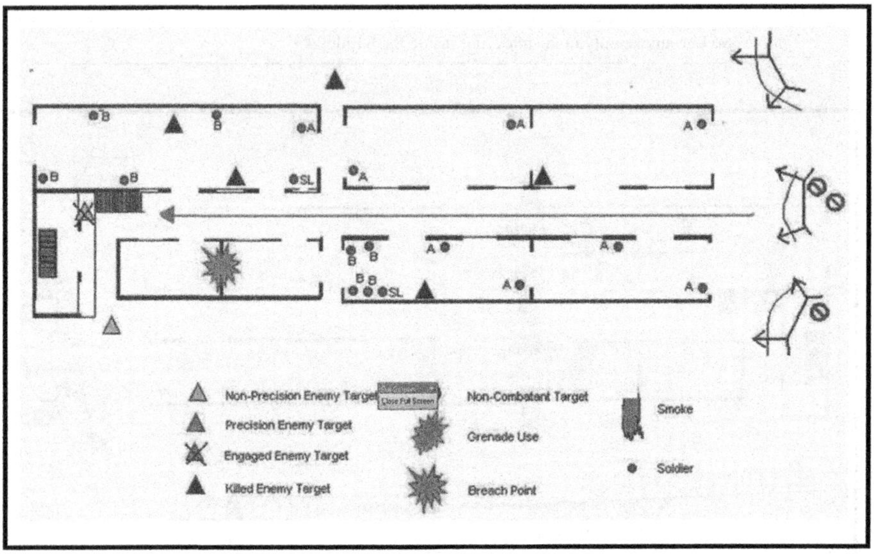

Figure 2-23. Obtaining obscurity.

2-62. While one team finds itself at the end of the building in their sector, the other team is faced with another wall and more clearing. One team prepares to breach the wall, and the other team takes up a hasty position and prepares for a counterattack. When the teams execute the breach, the next nonprecision target presents, and the support element engages it (Figure 2-24).

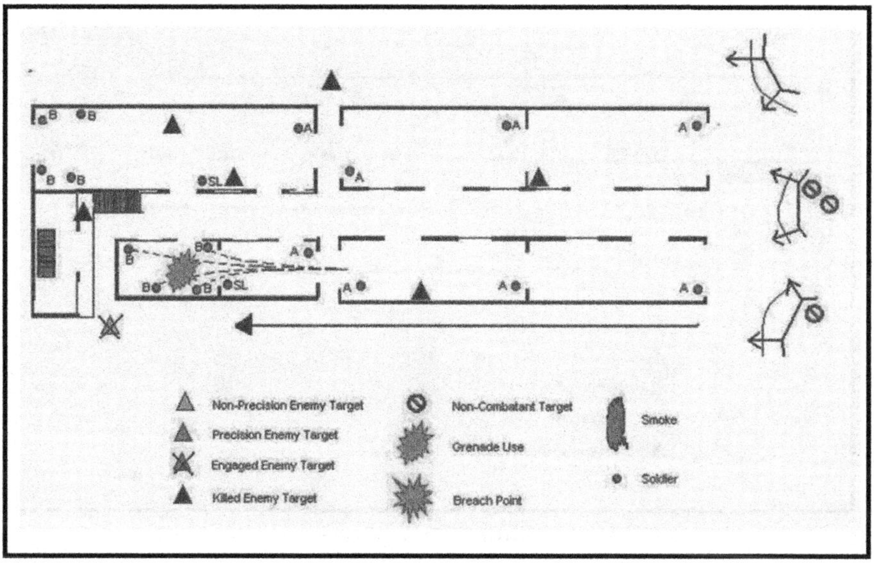

Figure 2-24. Clearing the wall.

2-63. Someone throws a grenade through the breach. The assault team enters and secures the room. When the grenade explodes inside the room, the final nonprecision target presents, and the support element engages it (Figure 2-25).

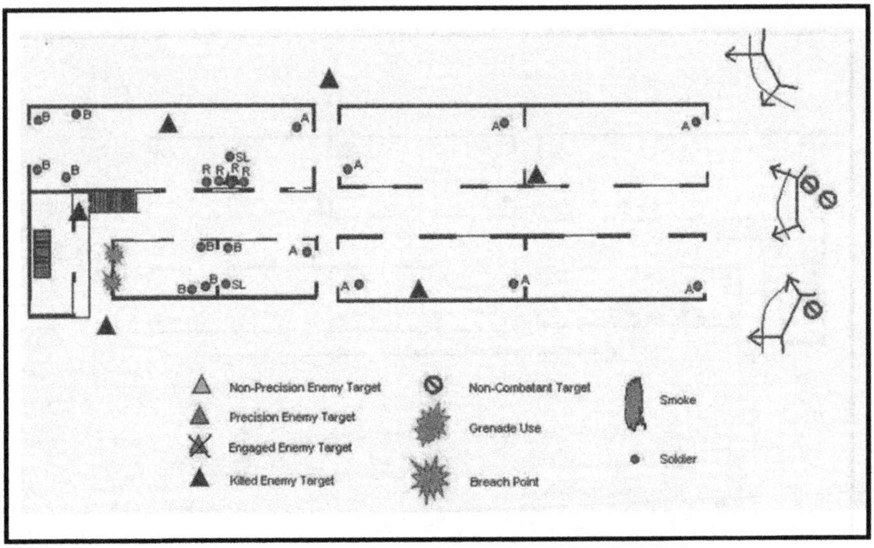

Figure 2-25. Securing a room.

2-64. With the first squad reaching its limit of advance, and with the second squad reaching the end of the building, the platoon leader calls another assault team forward from the reserve. The reserve assault team moves forward through the buildings, and positions itself to clear up the outside stairs of the two-story structure. The second squad places and detonates loophole charges on the far wall of the building. This supports the clearing of the last building (Figure 2-26).

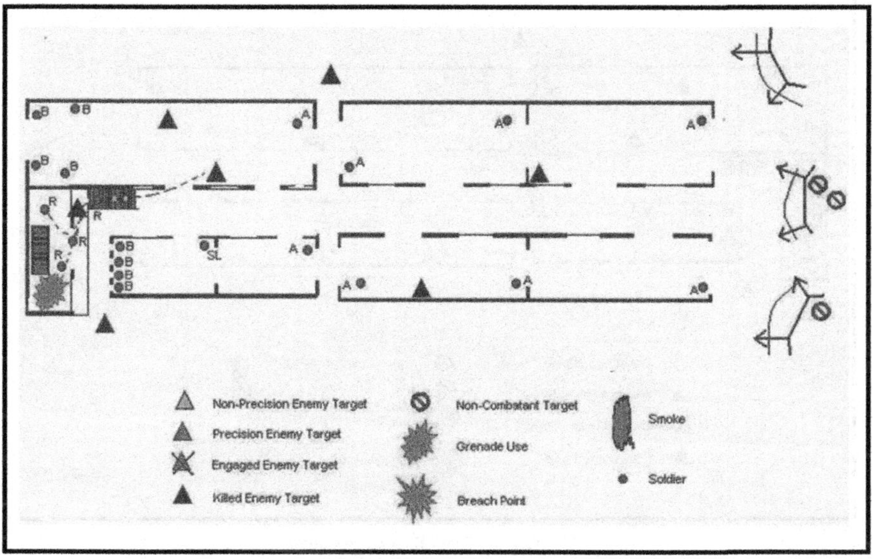

Figure 2-26. Reserve assault team moving forward.

2-65. Team B of the second squad moves into supporting fire positions. The new assault teams move up the outside stairs to the second floor of the final structure. They clear the balcony and the first room on the second floor. The support-by-fire position shift fires to the lower floor of the same structure (Figure 2-27).

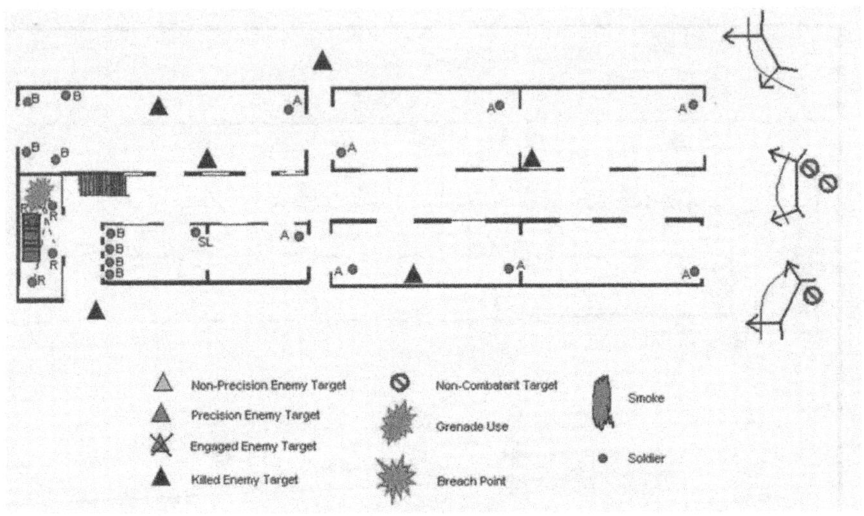

Figure 2-27. Team B, second squad--moving into the supporting fire position.

2-66. Once the floor is secure, the assault team moves down the stairs and clears the remaining floor (Figure 2-28). Meanwhile, the platoon repositions, and then consolidates and reorganizes.

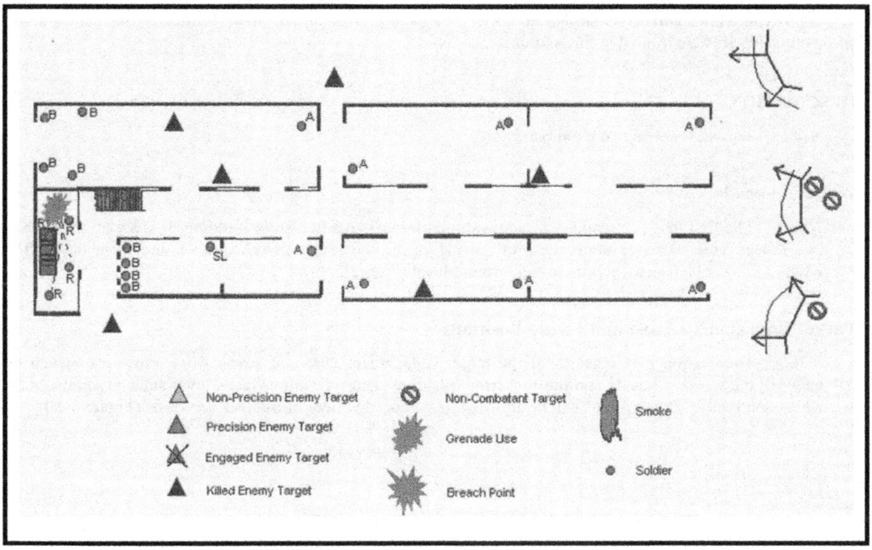

Figure 2-28. Consolidation and reorganization.

STATION 3: GRENADIER GUNNERY TRAINER

2-67. This is a live-fire station where M203 dual-purpose weapon gunners master the engagement of point-type M203 and M16 targets at 50 to 150 meters in an urban area. Tactical movement and fire commands are also trained at this station.

DESCRIPTION

2-68. This station has two main parts:

Wooden Facade

2-69. The first part is a wooden facade wall that looks like a two-story building. It is about 20 feet high by 30 feet wide. It has a doorway with outside steps, two lower floor windows and three upper floor windows, and two basement windows at ground level (Figure 2-29).

Target Engagement Lane and Firing Positions

2-70. The second part consists of the target engagement lane and seven firing positions, which are between the wooden façade and the first firing position. These positions have cover such as sandbags, log walls, and rubble. Firing points can be adjusted to support a variety of training scenarios (Figure 2-30).

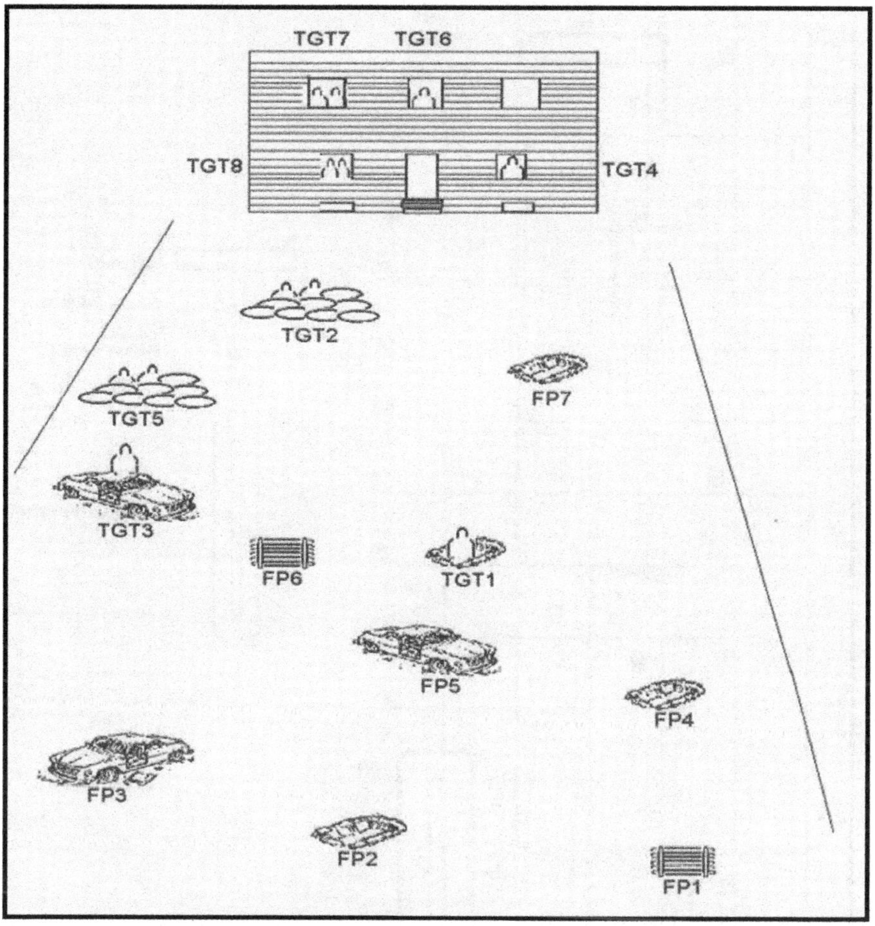

Figure 2-29. Part 1, Grenadier Gunnery Trainer, Station 3, wooden facade.

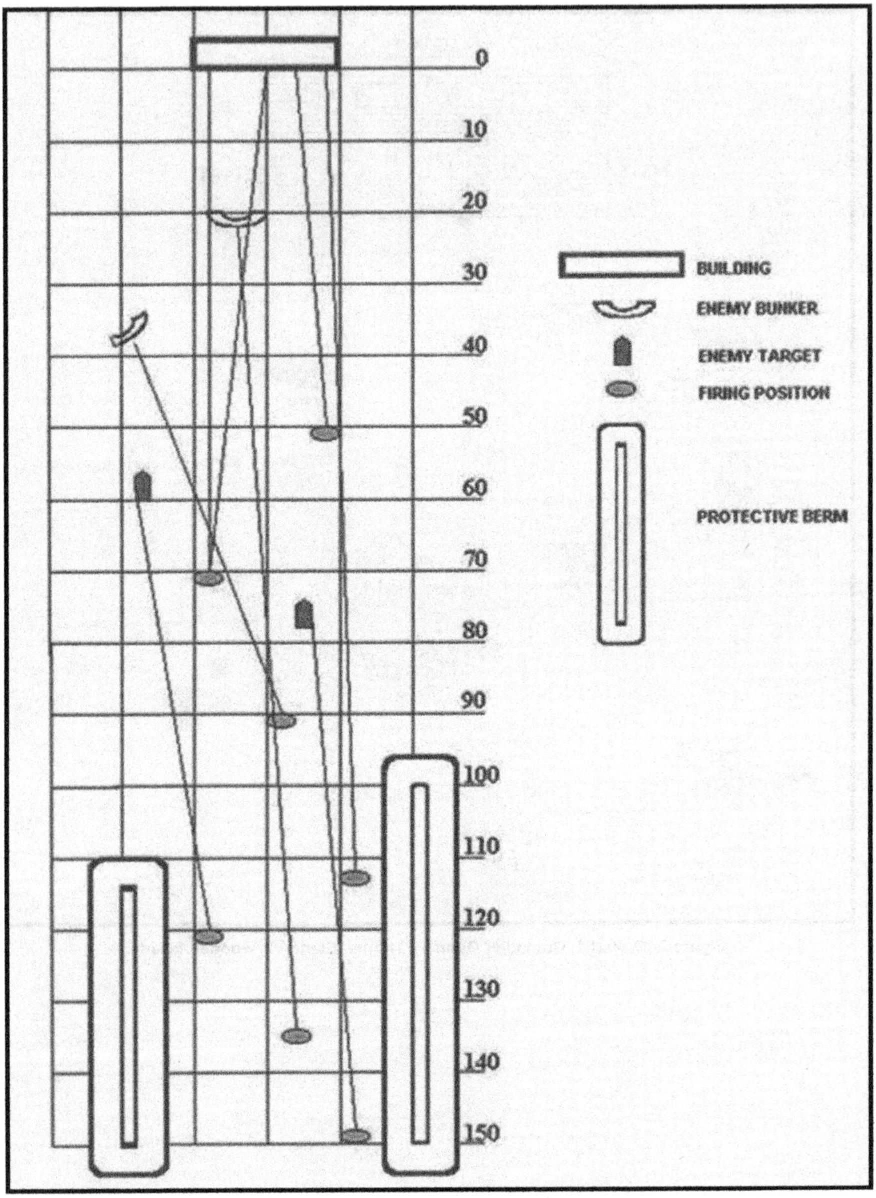

Figure 2-30. Part 2, Grenadier Gunnery Trainer, Station 3, target engagement lane and firing positions.

WARNING

Never engage targets with the 40-mm at ranges less than 31 meters, which is the minimum safe range. Trainers ensure the station accommodates this requirement as well as the 14-meter arming range.

TARGETRY

2-71. This station has twelve target mechanisms. These mechanisms are arranged so as to influence the gunner's decision as to which munition to use.

CONTROLLER REQUIREMENTS

2-72. As required by unit.

SAFETY REQUIREMENTS

2-73. Check the unit SOP and local safety regulations regarding live-fire ranges. Ensure all weapon systems are on SAFE before moving to the next firing position. Ensure that no 40-mm target is engaged at less than minimum arming range.

INDIVIDUAL AND COLLECTIVE TASKS TRAINED

2-74. These lists are noninclusive:

Supporting Individual Tasks

- 071-311-2007, *Engage Targets with M16-Series Rifle/M4 Carbine Rifle.*
- 071-311-2130, *Engage Targets with M203 Grenade Launcher.*
- 071-326-0503, *Move Over, Through, or Around Obstacles (except Minefields).*
- 071-326-0541, *Perform Movement Techniques during an Urban Operation.*
- 071-326-0557, *Select Hasty Firing Positions during an Urban Operation.*
- 071-326-0608, *Use Visual Signaling Techniques.*

Related Tactics, Techniques and Procedures

- FM 3-06.11, *Combined Arms Operations in Urban Terrain.*
  -- Chapter 3, *Offensive Operations.*
  -- Chapter 5, Fundamental Combat Skills.
- Appendix M, *Field Expedient Breaching of Common Urban Barriers.*

Training Support Requirements

2-75. Table 2-5 shows the minimum materials required to train at this station.

Table 2-5. Support requirements for Station 3.

Ammunition	*Rounds for each Weapon*	*Targets Required*
5.56-mm Ball	8	Four nonprecision human target mechanisms.
40-mm TP	8	Eight nonprecision human target mechanisms.

Conduct of Training

2-76. Whereas Table 2-5 showed the minimum amount of ammunition required for this station, Table 2-6 shows an example sequence of events and times, and Table 2-7 shows an example target scenario.

Table 2-6. Sequence of events and times for Station 3.

Event	*Action*	*Estimated Time*
1	*Perform Movement Techniques* *Select Hasty Firing Position* *Engage Targets With 5.56-mm (dry run)* *Engage targets with 40mm*	20 min
2	*Perform Movement Techniques* *Select Hasty Firing Position* *Engage Targets with 5.56mm (live-fire)* *Engage targets with 40mm*	10 min
NOTE: Training should include techniques outlined in FM 3-06.11		

Table 2-7. Example target scenario (presentation order) for Station 3.

Firing Position	*Weapon*	*Presentation Time*	*Range to Target*	*Target Presented*
1	5.56mm	6 sec	67 meters	Single Human Target (TGT 1)
2	40mm	2 min	105 meters	Multiple Human Targets (TGT 2)
3	5.56mm	6 sec	63 meters	Single Human Target (TGT 3)
4	5.56mm	6 sec	105 meters	Single Human Target (TGT 4)
5	40mm	2 min	57 meters	Multiple Human Targets (TGT 5)
6	40mm	2 min	70 meters	Multiple Human Targets (TGT 6)
6	5.56mm	6 sec	70 meters	Single Human Target (TGT 8)
7	40mm	2 min	50 meters	Multiple Human Targets (TGT 7)

STATION 4: URBAN OFFENSE/DEFENSE BUILDING, TASK/TECHNIQUE TRAINER

2-77. This station trains to attack or defend a building. Collective and individual TTP are trained. The platoon practices the task steps and performance measures for offensive and defensive tasks. This station increases the complexity of command, control, and maneuver (Figure 2-31).

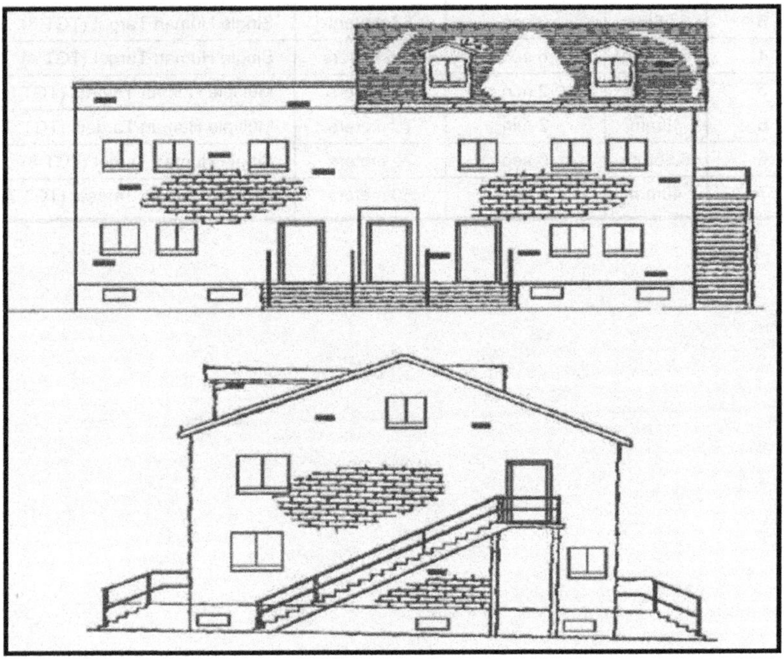

Figure 2-31. Urban offense/defense building, Station 4.

DESCRIPTION

2-78. This station is a two-story building with a basement. Each floor has several rooms, windows, doorways, loopholes, and mouseholes. The gabled roof has windows and a hatch to the attic below and half of the rooftop is flat. The building has exterior and interior stairways.

TARGETRY

2-79. This station has ten target mechanisms.

CONTROLLER REQUIREMENTS

2-80. These are based on unit requirements.

Safety Requirements

2-81. Only smoke, practice, and M84 stun grenades may be used at this station. It is not a live-fire station. Training is conducted using TES with blank 5.56-mm, or TES or SESAMS. Units should consult with their supporting combat engineers for the proper demolition effects simulators (DES) to use against the blow panels throughout the station.

DANGER

TO AVOID ASPHYXIATION, NEVER DETONATE SMOKE GRENADES INSIDE BUILDINGS. THEY ARE *ALL* DANGEROUS IN AN ENCLOSED SPACE.

WARNINGS

1. **To avoid injury, place your weapon on SAFE when climbing through obstacles such as windows, fences, and walls.**
2. **To avoid inflicting carbon burns or more serious injury, never fire blank ammunition at other Soldiers within 1 meter.**

Demolition Effects Simulators

2-82. Demolition effects simulators (DES) replicate breaching to gain entry to buildings and rooms. The walls of the UAC, Shoot House, and CACTF facilities may be breached using wall breach charge DES. Generally, the interior and exterior walls in these facilities are made with blow panels, plasterboard, or similar material, and measure about 4 feet square.

2-83. The DES charges for use at the UAC, shoot house, and the CACTF are constructed to breach a predetermined entry point without damaging the facility.

Field-Expedient Charges

- Flex-linear shaped charge.
- Doorknob charge.
- E-silhouette charge.

Operational Charges

2-84. Unit trainers must consult with installation safety and combat engineers to determine the appropriate reduced charge. See also FM 3-06.11.

Individual or Collective Tasks Trained

Collective Task

- FM 3-21.8 (7-8).
  -- 07-3-9018, *Clear a Building (Platoon/Squad).*
  -- 07-6-1216, *Defend an Urban Area (Battalion/BCT).*

Related Collective Tasks

- 07-2-5027, *Conduct Consolidation and Reorganization (Company/Platoon).*
- 08-2-0003. *Treat Casualties.*
- 08-2-0004. *Evacuate Casualties.*
- 19-3-3100, *Process Detainees at a Collecting Point (CP).*
- 07-2-1342, *Conduct Tactical Movement (Company/Platoon).*
- 07-2-1189, *Conduct a Dismounted Tactical Road March (Company/Platoon).*
- 07-2-1198, *Conduct a Mounted Tactical Road March (Company/Platoon).*
- Warrior Battle Drill 12, *Enter and Clear a Room.*

Supporting Individual Tasks

- 052-193-1013, *Neutralize Booby Traps.*
- 071-010-0006, *Engage Targets with M249 Machine Gun.*
- 071-311-2007, *Engage Targets with M16-Series Rifle/M4 Carbine Rifle.*
- 071-315-2308, *Engage Targets with M16-Series Rifle/M4 Carbine Using a Night Vision Sight AN/PVS-4.*
- 071-325-4407, *Employ Hand Grenades.*
- 071-326-0501, *Move as a Member of a Fire Team.*
- 071-326-0503, *Move Over, Through, or Around Obstacles (except Minefields).*
- 071-326-0541, *Perform Movement Techniques during an Urban Operation.*
- 071-326-0557, *Select Hasty Firing Positions during an Urban Operation.*
- 071-326-0608, *Use Visual Signaling Techniques.*
- 071-326-5605, *Control Movement of a Fire Team.*
- 071-326-5611, *Conduct the Maneuver of a Squad.*
- 071-710-0008, *Operate Night Vision Goggles AN/PVS-7B.*
- 113-571-1022, *Perform Voice Communications.*
- 191-377-5250, *Handle Enemy Personnel and Equipment.*

Related Tactics, Techniques, and Procedures

- FM 3-06.11, *Combined Arms Operations in Urban Terrain.*
  - -- Chapter 3, *Offensive Operations.*
  - -- Chapter 5, *Fundamental Combat Skills.*
  - -- Appendix M, *Field Expedient Breaching of Common Urban Barriers*
- FM 3-22.9, *Rifle Marksmanship.*

Training Support Requirements

2-85. Table 2-8 shows the minimum ammunition required to train at this station.

Table 2-8. Ammunition requirements for Station 4.

Ammunition	Rounds for each Weapon
5.56-mm Blank	16
5.55-mm Blank (Linked)	24
7.62-mm Blank	100
Practice Grenade Bodies	5
Practice Grenade Fuzes	10 per platoon
M84 Stun Grenade Bodies	5
M84 Stun Grenade Fuzes	10 per platoon
Smoke Grenades	4 per platoon

Conduct of Training

2-86. Table 2-9 shows the sequence of training for Station 4. Figures 2-32 and 2-33 show example scenarios of training events that can be conducted at this station. These are only examples. The unit should develop training events based on its METL and training needs.

Table 2-9. Sequence of events and times for Station 4.

Event	Action	Estimated Time
1	*Clear a Building*	2 Hours Crawl
		1 Hour Walk
		30 Minutes Run
2	*Defend Built-up Area/Building*	1 Hours Crawl
		1 Hour Walk
		2. Hour Run
3	*Conduct AAR*	1 Hour

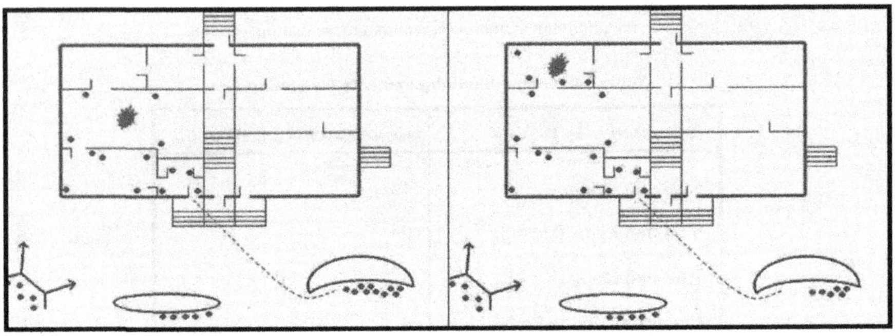

Figure 2-32. Event 1, *Clear a Building.*

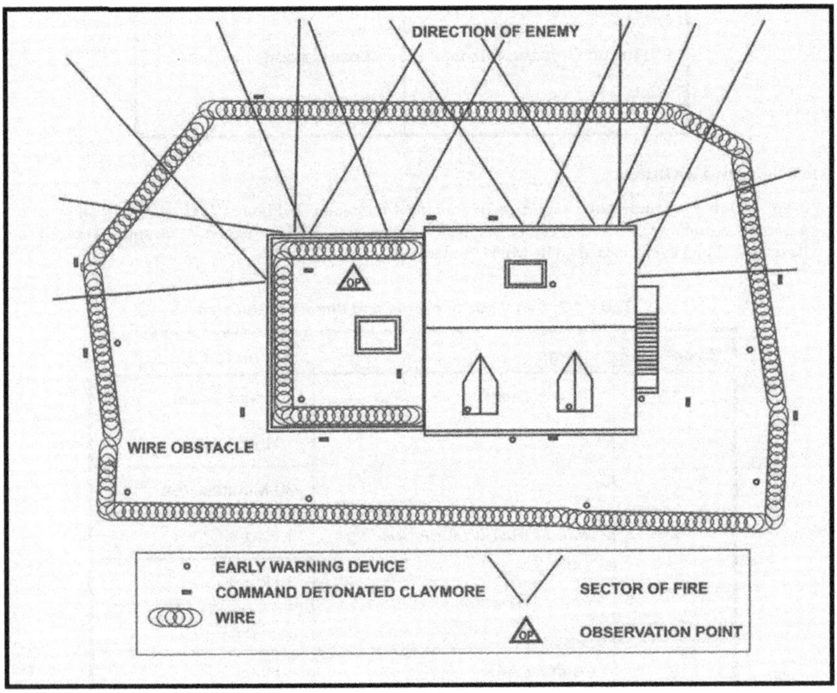

Figure 2-33. Event 2, *Defend a Building.*

STATION 5: UNDERGROUND TRAINER

2-87. This station provides training up to squad level for subterranean operations, specifically, the task *Clear and Move Tactically in a Subterranean Environment.*

DESCRIPTION

2-88. This station is a sewer system built in an irregular pattern, with four manhole covers. Two access tunnels lead from the outside tunnel to the underground loop, and the station is completely covered by dirt, only the manholes and ground level entrances are exposed (Figure 2-34).

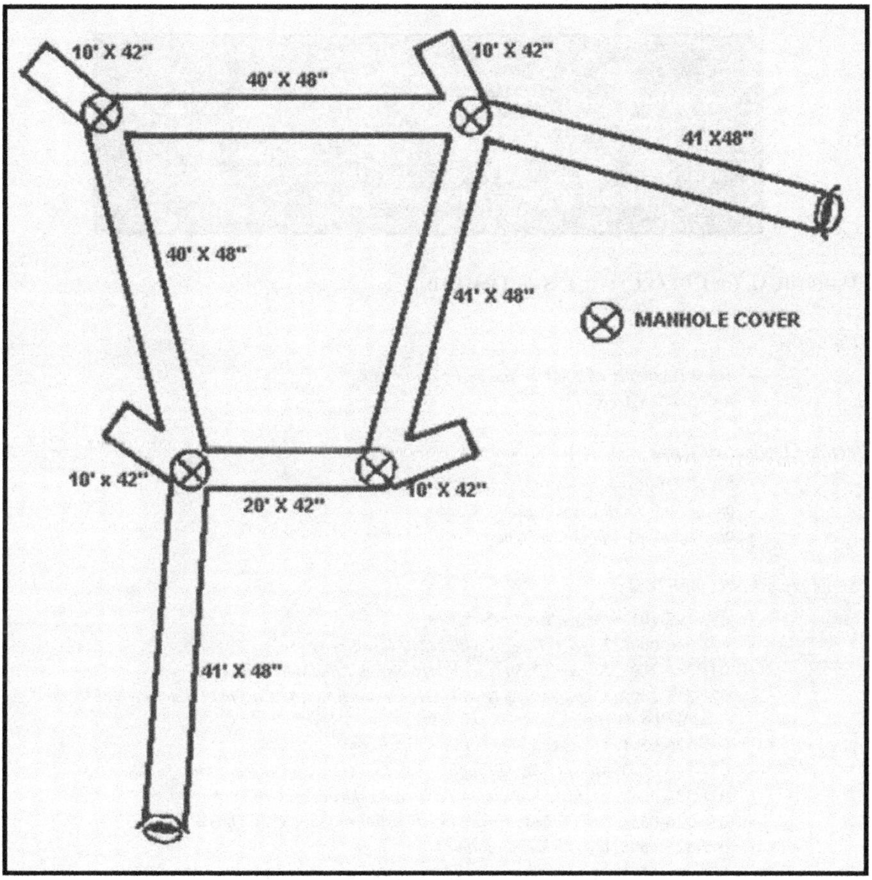

Figure 2-34. Underground Trainer, Station 5.

TARGETRY

2-89. Station 5 is a force-on-force training station and has no targetry.

CONTROLLER REQUIREMENTS

2-90. As required by unit.

SAFETY REQUIREMENTS

2-91. Soldiers should wear their helmets at all times when in the sewer system to avoid head injuries. Smoke grenades or M84 stun grenades *will not* be used in the underground trainer.

> **DANGER**
>
> TO AVOID ASPHYXIATION WHEN IN THE TRAINER, NEVER DETONATE SMOKE GRENADES IN THE SEWER. USING ANY TYPE OF EXPLOSIVE SIMULATOR IS DANGEROUS IN AN ENCLOSED SPACE. THE PROTECTIVE MASK IS NOT EFFECTIVE IN OXYGEN-DEFICIENT ATMOSPHERES.

INDIVIDUAL OR COLLECTIVE TASKS TRAINED

Collective Task

- *Move Tactically in a Subterranean Environment.*
- *Reconnoiter Area.*

Related Collective Tasks

- 07-2-1270, *Conduct an Infiltration (Company/Platoon).*
- 08-2-0003. *Treat Casualties.*
- 08-2-0004. *Evacuate Casualties.*

Supporting Individual Tasks

- 052-193-1013, *Neutralize Booby Traps.*
- 071-010-0006, *Engage Targets with M249 Machine Gun.*
- 071-311-2007, *Engage Targets with M16-Series Rifle/M4 Carbine Rifle.*
- 071-315-2308, *Engage Targets with M16-Series Rifle/M4 Carbine Using a Night Vision Sight AN/PVS-4.*
- 071-326-0501, *Move as a Member of a Fire Team.*
- 071-326-0503, *Move Over, Through, or Around Obstacles (except Minefields).*
- 071-326-0541, *Perform Movement Techniques during an Urban Operation.*
- 071-326-0557, *Select Hasty Firing Positions during an Urban Operation.*
- 071-326-0608, *Use Visual Signaling Techniques.*
- 071-326-5605, *Control Movement of a Fire Team.*
- 071-326-5611, *Conduct the Maneuver of a Squad.*
- 071-710-0008, *Operate Night Vision Goggles AN/PVS-7B.*

- 113-571-1022, *Perform Voice Communications.*
- 191-377-5250, *Handle Enemy Personnel and Equipment.*

Related Tactics, Techniques and Procedure

- FM 3-06.11, *Combined Arms Operations in Urban Terrain.*
  -- Chapter 3, *Offensive Operations.*
  -- Chapter 5, *Fundamental Combat Skills.*
  -- Appendix M, Field Expedient Breaching of Common Urban Barriers.
- FM 3-22.9, *Rifle Marksmanship.*

TRAINING SUPPORT REQUIREMENTS

2-92. Use Table 2-10 for the minimum ammunition required to train at this station.

Table 2-10. Ammunition requirements for Station 5.

Ammunition	Rounds for each Weapon
5.56-mm Blank	20
5.55-mm Blank (Linked)	40

CONDUCT OF TRAINING

2-93. Table 2-11 shows the sequence of training for this station. These are only examples. The unit should develop training events based on its METL and training needs.

Table 2-11. Sequence of events and times for Station 5.

Event	Action	Estimated Time
1	*Reconnoiter*	30 minutes (Crawl) 30 minutes (Walk) 30 minutes (Run)
2	*Conduct an Infiltration (Co/Plt)*	30minutes (Crawl) 30 minutes (Walk) 30 minutes (Run)
3	*Conduct AAR*	30 minutes

Chapter 3
Shoot House

This chapter describes the live-fire Shoot House and provides layouts and training scenarios. Fighting in built-up areas is fragmented; gains are often small. Therefore, units depend on the initiative, skill, and discipline of small-unit leaders and individual Soldiers. Leaders must be competent and confident in their urban operations skills and must have the courage to accomplish their missions while isolated from their parent units. Soldiers and leaders require mission-type orders that allow for decentralized execution.

PURPOSE

3-1. The Shoot House provides the leader with a facility to train and evaluate the unit during a live-fire exercise. Units are trained and evaluated on their ability to conduct tactical movement, engage targets, conduct breaches, and practice target discrimination in an urban environment.

DESCRIPTION

3-2. This facility has two parts, the live-fire facility and the AAR room:

LIVE-FIRE FACILITY

3-1. The live-fire facility is a single-level structure with a barn roof, catwalk, and eight adjoining rooms. Each room has video capability for AARs. The walls are bulletproof and prevent ricochets. Each side has an entrance. Replaceable blow panels throughout the facility can withstand reduced DES blasts. These also offer alternative entry points. A crane system in the roof allows trainers to selectively replace doorways and blow panels with solid wall sections. Before training explosive and mechanical breaching operations, trainers should place mobile exterior baffles or containment walls placed by each entry to prevent rounds and demolition effects from escaping. All of these capabilities let the unit vary its training scenarios. For safety, the Shoot House is windowless (Figure 3-1).

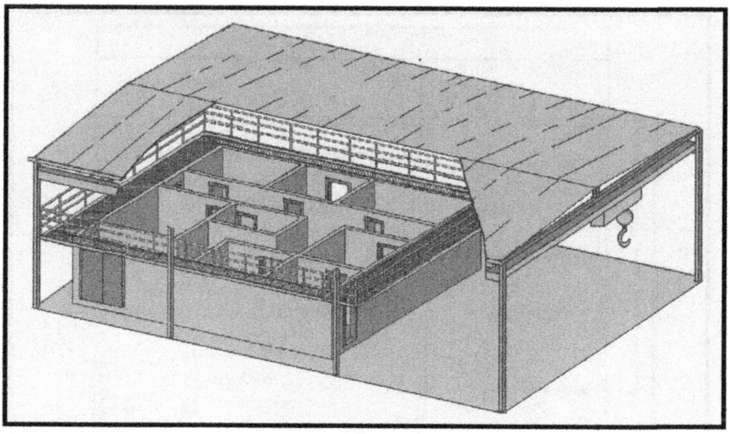

Figure 3-1. Shoot House.

AAR BUILDING

3-2. The AAR building has a conference/AAR room that can accommodate up to a platoon size element, an instrumentation and targetry control room, and an AAR edit and projection room that can handle a formal AAR (Figure 3-2).

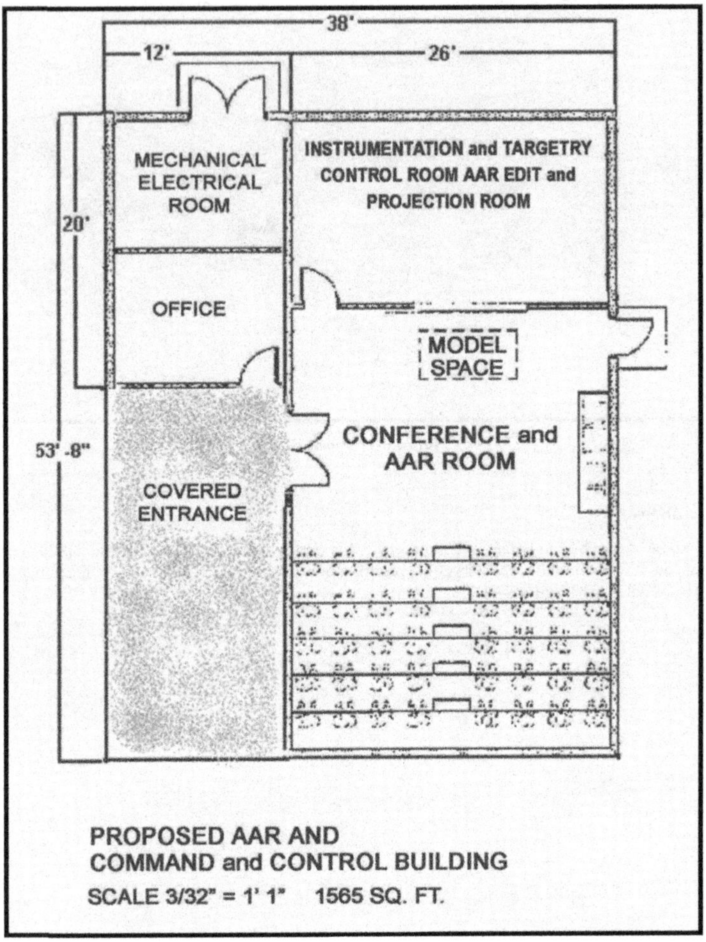

Figure 3-2. Command and control building and AAR room.

TRAINING AUDIENCE

3-3. The facility provides leaders the ability to train and evaluate the unit during a live-fire exercise. However, the unit's training requirements determine the size of the unit and the tasks to be trained at this facility. The unit develops its own organization and the steps for conducting the exercise based on its METL and training needs.

TRAINING INFORMATION

3-4. Tactical operations in an urban environment may range from combat missions to stability and support missions. Commanders should note that the Shoot House *is not* suited for the employment of live-fragmentation/concussion grenades, 40-mm HE/HEDP, antiarmor weapons, mortars, artillery, or air delivered munitions. Crew-served weapon systems, M2 and M1 tanks, and machine guns can be employed against supplemental targetry that is exterior to the Shoot House. The Shoot House has specially constructed areas that are used to conduct either mechanical or explosive breaching techniques. When conducting explosive breaching, units should consult with local Engineers for appropriate DES charges and consult Appendix A before conducting the training.

CONDUCT OF TRAINING

3-5. The unit develops realistic time frames for each operation or task. The times are based on the *crawl-walk-run* concept. The training event or STX can be arranged in any order as long as they follow a tactical scenario. *Assault a Building*, *Enter a Building and Clear a Room*, and *Clear a Building* are the primary collective tasks trained at this facility.

EXECUTION

3-6. Tables 3-1 through 3-5, pages 3-4 through 3-7, shows a sequence of events and times, and Figure 3-3 through Figure 3-6, pages 3-6 through 3-9, show example scenarios. These are only examples. The unit should develop training events based on its METL and training needs. However, the leader should do the following when conducting training at the live-fire Shoot House:

1. Identify the unit's critical missions (see commander's training guidance), and select those tasks that support specific tasks related to urban operations.
2. Arrange the related tasks in the order of their importance to mission accomplishment.
3. Assess the unit's current level of proficiency (strengths and weaknesses), and determine the degree of difficulty of each task. The training events and scenarios should start from the least difficult to the most difficult tasks. This gives the Soldiers confidence.
4. Identify the individual and leader tasks to be trained.
5. Conduct individual training first, because it is the first step in the building-block approach. It is also the basis for unit or collective task training. Mastery of individual skills is a precondition for effective training.
6. Set up training conditions based on unit proficiency. At first, keep the conditions basic, until Soldiers attain proficiency. Increase difficulty and realism as training progresses. Then, add limited visibility, CBRN factors, and live fires with service or short-range training ammunition (SRTA).
7. Understand that the Shoot House is a live-fire capable training facility. However, it can be used to train any level of proficiency, including blank, TES, SESAM, SRTA, and ball (no tracers). The leader assesses the unit's proficiency.

Table 3-1. Sequence of events and times for the platoon.

Element	*Event action*	*Estimated Time*
All 1st Squad 2d Squad 3rd Squad	1. Conduct precombat checks (for example; install, align, and troubleshoot MILES equipment; conduct inspection; draw equipment). 2. Support Operations. 3. Conduct Rehearsals. 4. Conduct Event 1 (Crawl).	2 Hours
3rd Squad 1st Squad 2d Squad	5. Support Operations. 6. Conduct Rehearsals. 7. Conduct Event 1 (Crawl).	2 Hours
2d Squad 3rd Squad 1st Squad	8. Support Operations. 9. Conduct Rehearsals. 10. Conduct Event 1 (Crawl).	2 Hours
1st Squad 2d Squad 3rd Squad	11. Support Operations. 12. Conduct Rehearsals. 13. Conduct Event 1 (Walk).	2 Hours
3rd Squad 1st Squad 2d Squad	14. Support Operations. 15. Conduct Rehearsals. 16. Conduct Event 1 (walk).	2 Hours
2d Squad 3rd Squad 1st Squad	17. Support Operations. 18. Conduct Rehearsals. 19. Conduct Event 1 (walk).	2 Hours
1st Squad 2d Squad 3rd Squad	20. Support Operations. 21. Conduct Rehearsals. 22. Conduct Event 1 (Run).	1 Hours
1st Squad 2d Squad 3rd Squad	23. Support Operations. 24. Conduct Rehearsals. 25. Conduct Event 1 (Run).	1 Hours
1st Squad 2d Squad	26. Support Operations. 27. Conduct Event 1 (Run).	1 Hours

Table 3-2. Event 1, Squad Sequence (*Crawl*).

Element	*Event Action*	*Estimated Time*
1 (CRAWL)	**Terminal Learning Objective** Familiarize the squad with how to enter and clear a room during the crawl phase of training **Tasks** *Enter and Clear a Room (Squad)* • *Perform Movement Technique* • *Select Hasty Firing Position* *Conduct a Breach (Mechanical)* • *Employ Hand Grenades* • *Engage Targets with M16A2/M4* • *Engage Targets with M249 Machine Gun*	**2 Hours**
NOTES: 1. Perform controlled pairs and target discrimination techniques (FM 3-22.9). 2. Include techniques (FM 3-06.11) and performance measures (FM 3-21.8 and DTMS).		

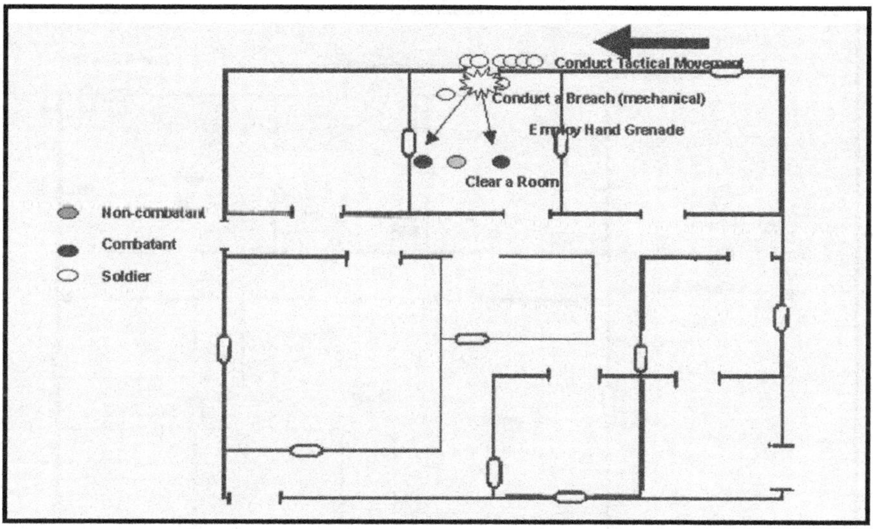

Figure 3-3. Event 1, *Squad Scenario (Crawl)*, force-on-targetry.

Table 3-3. Event 2, Squad Sequence (*Walk*).

Element	*Event Action*	*Estimated Time*
2 (WALK)	**Terminal Learning Objective** Familiarize the squad with how to enter and clear multiple rooms during the walk phase of training **Tasks** *Enter and Clear a Building (Squad)* • *Perform Movement Technique* • *Select Hasty Firing Position* *Conduct a Breach (Explosive DES)* • *Engage Targets with M16A2/M4* • *Engage Targets with M249 Machine Gun* *Conduct AAR*	2 Hours
NOTES: 1. Perform controlled pairs and target discrimination techniques (FM 3-22.9) 2. Include techniques (FM 3-06.11) and performance measures (FM 3-21.8 and DTMS).		

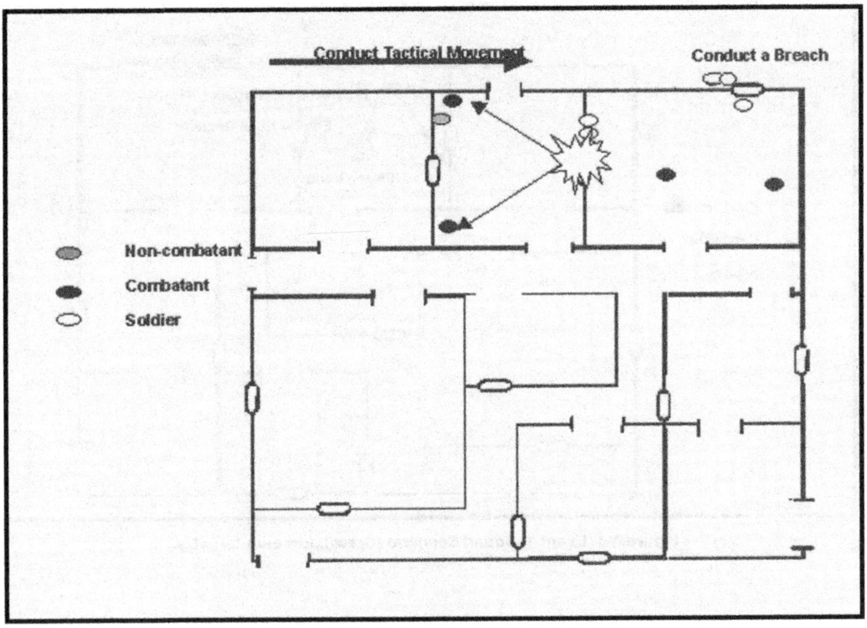

Figure 3-4. Event 2, *Squad Scenario (Walk)*, force-on-force.

Table 3-4. Event 3, Squad Sequence (*Run*).

Element	*Event Action*	*Estimated Time*
3 (RUN)	**Terminal Learning Objective** Refamiliarize squad with how to enter and clear multiple rooms during the run phase of training. **Tasks** *Enter and Clear a Building (Squad)* • *Perform Movement Technique* • *Select Hasty Firing Position* *Conduct a Breach (Mechanical)* • *Engage Targets with M16A2/M4* • *Engage Targets with M249 Machine Gun* *Clear a Hallway* • *Engage Targets with M16A2/M4* • *Engage Targets with M249 Machine Gun* *Conduct a Breach (explosive DES)* • *Engage Targets with M16A2/M4* • *Engage Targets with M249 Machine Gun* *Conduct AAR*	**1 Hour**

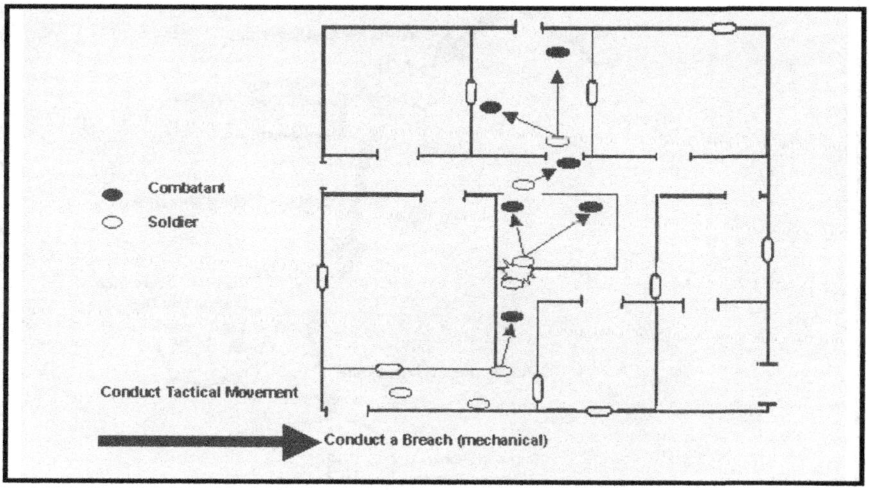

Figure 3-5. Event 3, *Squad Scenario (Run)*, force-on-targetry.

Table 3-5. Event 4, Platoon Evaluation Sequence (*Run*).

Element	*Event Action*	*Estimated Time*
4 (RUN)	*Occupy Assembly Area* *Develop and Communicate a Plan* *Prepare for Combat* *Perform Passage of Line* *Conduct Tactical Movement* *Break Contact* *Conduct Intermediate AAR* *Conduct Tactical Movement* *Perform Overwatch/Support by Fire* *Assault a Building* *Consolidate and Reorganize* *Conduct Final AAR*	1 Hour 2 Hours 30 min 1 Hour 5 min 20 min 1 Hours 1 Hours 2 Hours 1 Hours

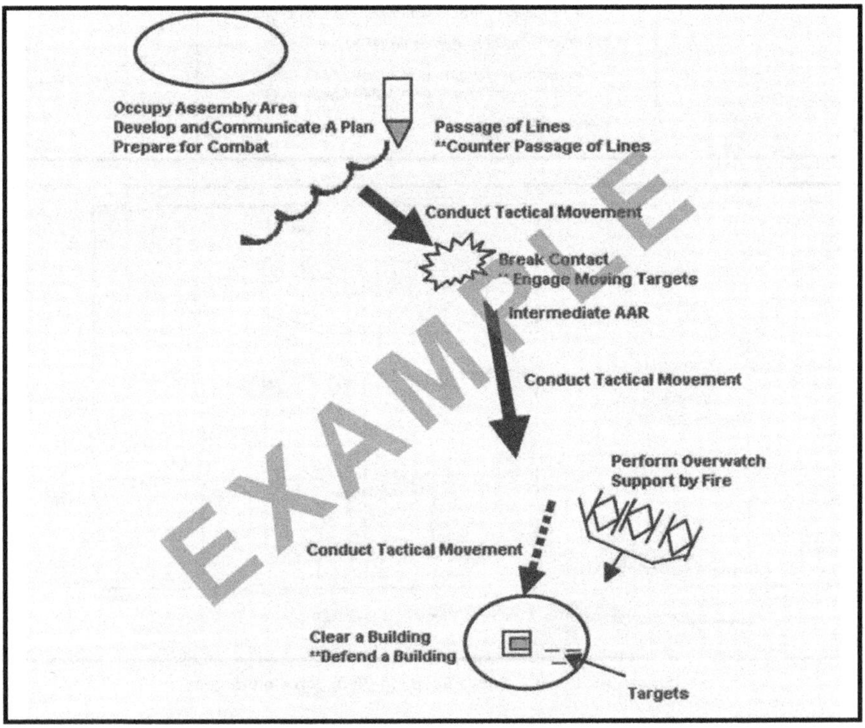

Figure 3-6. Example platoon tactical scenario (*Run*).

Operation Order

3-7. Figure 3-7 shows an example OPORD for the conduct of training.

```
                          ______________
                         (Classification)
                                          Copy ____ of ___ copies
                                        ____Company _____ Infantry
                                       (Unit Location)____________
                                        Date/time Group __________

(Message Reference Number)

OPORD: _____________
Reference Map, Series ___________________
Time Zone used Throughout the Order:_____________________

TASK ORGANIZATION:

______Platoon Platoon
______Platoon ____ 60-mm Section
______Antiarmor Section

1. SITUATION

   a. Enemy Forces. Annex B (Intelligence Overlay [prepared by
battalion S-2]).

      (1) The company is opposed by elements of the Red Company of the
Blue Brigade at about 60 percent strength. Within our zone, we have a
platoon-sized element occupying a defensive position in a building in
the vicinity of (______). The enemy has small arms weapons and may
call on company mortars for support.

      (2) Current indications are the enemy is regrouping to initiate
small-unit offensive actions in the area. His most likely course of
action is to move to the (direction) to link up with elements of the
parent unit.

   b. Friendly Forces.

      (1) (___) Infantry attacks at (______) hours to seize Objective
ALPHA (grid). The intent is to prevent the enemy element from moving
(direction) to link up with other enemy units.
      (2) Company (__) on our (left, right) makes the main attack at
(____) hours to seize Objective RED (grid) in order to establish
blocking positions to stop the enemy's movement to the (direction).

                          ______________
                         (Classification)
```

Figure 3-7. Example company OPORD.

(Classification)

2. MISSION

Company (__) attacks to seizes Objective BLUE (grid) at (date/time) hours in order to prevent the enemy from regrouping and engaging Company (__) (main effort) from Objective BLUE.

3. EXECUTION

a. **Concept of Operation**. Annex C (Operations Overlay). (This should include those steps listed in the task , Conduct Troop-Leading Procedures, paragraph 17. These should be written out when possible and not just placed on an overlay.)

(1) Maneuver. (___) Platoon establishes a defense in its zone, NLT (date/time) hours, to destroy small enemy elements moving or exfiltrating to the (direction), from the vicinity of Objective BLUE. This defense supports the attack on Objective BLUE by completing destruction of any enemy that may escape. (Evaluated) Platoon conducts the company main attack at (hours) to seize Objective BLUE to destroy the enemy or force them to withdraw into (___) Platoon's defense. (___) Platoon supports the attack of the (Evaluated) Platoon on Objective BLUE to suppress enemy fires and isolate the objective. On seizure of Objective BLUE, (Evaluated) Platoon and (___) Platoon will defend in sector to destroy the enemy elements withdrawing to avoid (___) Company's attack.

(2) **Fire support**.

(a) Priority of fires.

- Initially to (Evaluated) Platoon.
- Upon seizure of Objective BLUE, priority shifts to (Defensive) Platoon.

(b) Priority targets.

- 60-mm priority target initially AA4031.
- On initiation of assault on Objective BLUE, priority shifts to target AA4027 to isolate the objective.
- Upon seizure of Objective BLUE, priority shifts to target AA4025.

(Classification)

Figure 3-7. Example company OPORD (continued).

(Classification)

b. **Tasks to maneuver units.**

(1) (___) Platoon. Destroy enemy elements moving to the (direction), NLT (date/time).

(2) (Evaluated) Platoon. Consolidate and defend in sector to destroy enemy elements following the seizure of Objective BLUE.

(3) (___) Platoon. Be prepared to assault Objective BLUE, on order, consolidate and defend in sector to destroy enemy elements following the seizure of Objective BLUE.

c. **Tasks to Subordinate Units.**

(1) 60-mm Section. Initially in the vicinity of CP 7, priority of fires to the (Evaluated) Platoon. Move with the (Supporting) Platoon. On order, priority of fires to the (Defensive) Platoon.

(2) Antiarmor Section. Attached to the (Supporting) Platoon.

d. **Coordinating Instructions:**

(1) (Defensive) Platoon reports defense established and locations.
(2) Platoon reports the seizure of all objectives.
(3) Platoon reports the consolidation and establishment of the defense.
(4) Order of march from the assembly area to the objective: defense platoon, evaluated platoon, headquarters, support platoon, 60-mm mortar.

4. **SERVICE SUPPORT**

a. **General.** Company trains collocated with the battalion combat trains in the vicinity of (grid). Supply distribution at the company CP.

(Classification)

Figure 3-7. Example company OPORD (continued).

(Classification)

b. **Materiel and Services.**

(1) Class I: Ration cycle MRE-MRE-MRE.

(2) Class V: Pick up basic load at company CP, NLT (date/time).

(3) Medical: Company casualty collection point is __________.

(4) EPW: EPW collection point is ___________.

5. **COMMAND AND SIGNAL**

a. **Command.** Company command group initially follows the (Evaluated) Platoon.

b. **Signal.**

(1) Radio-listening silence is in effect until contact is made.

(2) SOI _____ is in effect.

JONES
Commanding

AnnexES:
B-Intelligence Overlay
C-Operations Overlay
D-Fire Support Overlay

(Classification)

Figure 3-7. Example company OPORD (continued).

TRAINING SUPPORT REQUIREMENTS

3-8. Minimum trainers and evaluators should be based on unit proficiency, training needs, and local safety requirements. Table 3-6 contains consolidated support requirements.

Table 3-6. Consolidated support requirements.

Ammunition	*Event 1*	*Event 2*	*Event 3*	*Event 4 (Optional)*
5.56-mm (per weapon) ball, blank, SRTA, or SESAMS	20 Blank	20 SESAMS	20 Ball/SRTA	20 Ball/SRTA
5.56-mm linked (per weapon)	30 Blank	30 SESAMS	30 Ball/SRTA	30 Ball/SRTA
Hand grenade body (practice)	0	--	4	4
Hand grenade fuzes (practice) per platoon	0	--	4	4
Hand grenade, smoke, per platoon	0	6	6	6
M84 hand grenade, stun, per platoon	0	4	4	0
TES (MILES Equipment)				
M16 (1 per weapon, 1 per OPFOR)	NO	YES	NO	NO
M249 (1 per weapon)	NO	YES	NO	NO
SAAF (1 per platoon)	NO	YES	NO	NO
Controller Gun (1 per OC)	NO	YES	NO	NO
Targetry	5	0	5	10
OPFOR	NO	6	NO	NO

TRAINING AND EVALUATION

3-9. The T&EOs are the foundation of the MTP and the collective training of the platoon. T&EOs are training objectives (task, conditions, and standards) for the collective tasks that support platoon critical wartime operations. T&EOs may be trained separately, in an STX, in an FTX, or in live-fire exercises. For collective live-fire standards, the trainer needs to refer to the applicable Infantry Gunnery Manual for the appropriate course of fire. Those standards and courses of fire need to be integrated into the training exercise. Table 3-8 provides a list of related collective and supporting individual tasks.

Table 3-7. Related collective and supporting individual tasks.

Task	FM 3-21.8	FM 3-06.11	STP 21-1-SMCT	STP 7-11B1-SM-TB	STP 21-24-SMCT
Conduct Consolidation and Reorganization Treat Casualties Evacuate Casualties Process Enemy Prisoners of War/Captured Materiel Defend MOUT Building Conduct Tactical Movement	X	X			
Supporting Individual Tasks					
Engage Targets with an M16-Series Rifle			071-311-2007		
Engage Targets with an M16-Series Rifle Using a Night Vision Sight AN/PVS-4			071-315-2007		
Engage Targets with an M249 Machine Gun				071-010-0006	
Employ Hand Grenades			071-325-4407		
Move Over, Through, or Around Obstacles (Except Minefields)			071-326-0503		
Neutralize Booby Traps					052-193-1013
Move as a Member of a Fire Team				071-326-0501	
Perform Movement Techniques during MOUT		X		071-326-0541	
Conduct the Maneuver of a Squad				071-326-5611	
Conduct an Explosive Breach				071-440-0030	
Execute a Mechanical Breach				071-440-0027	
Conduct a Breach				071-440-0029	
Operate Night Vision Goggles AN/PVS-7				071-710-0008	
Select Hasty Firing Positions during MOUT				071-326-0557	
Control Movement of a Fire Team				071-326-5605	

Chapter 4

Breach Facility

Urban operations require Soldiers to enter buildings through walls, doors, windows, and roofs. The Breach Facility is designed to train Soldiers using tasks and techniques on how to breach locked doors, windows, and create man-sized holes in walls. The facility can be used to train mechanical, ballistic, thermal, and explosive breaching.

PURPOSE

4-1. The Breach Facility (Figure 4-1) is used to train Soldiers semiannually on the technical aspects of breaching techniques. It is also used to train TTP and explosive techniques not trained on any other type range.

Figure 4-1. Breach facility.

DESCRIPTION

4-2. This facility has three stations:

STATION 1

4-3. This station is a facade that is 8 feet high and 88 feet wide with four sections of two doors for a total of eight doors. Soldiers can conduct mechanical, thermal, ballistic, and explosive breaching techniques on doors at this station. Half of the doors open from the exterior, and the other half of the doors open from the interior (Figure 4-2).

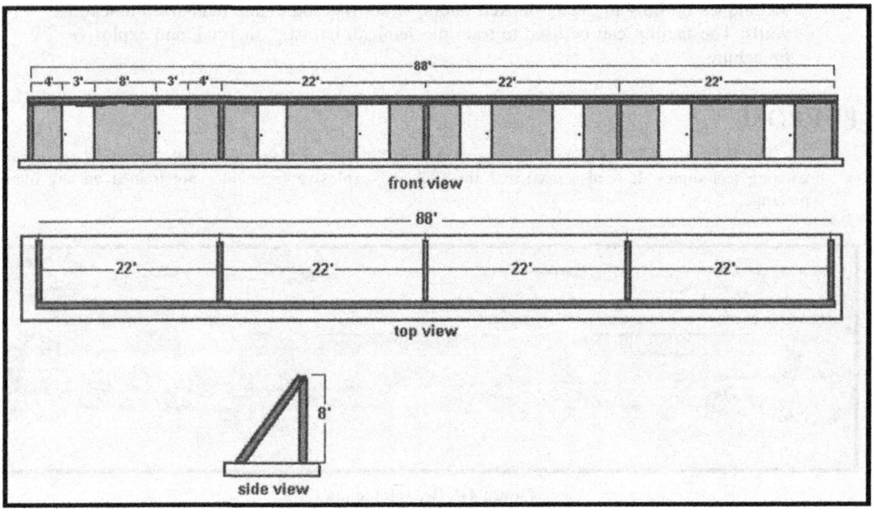

Figure 4-2. Station 1—Door breaching.

STATION 2

4-4. This station is a facade. It measures 10 feet high by 88 feet wide. It has four sections of two windows each, for a total of eight windows. Soldiers can conduct mechanical and explosive breaching techniques on windows at this station (Figure 4-3).

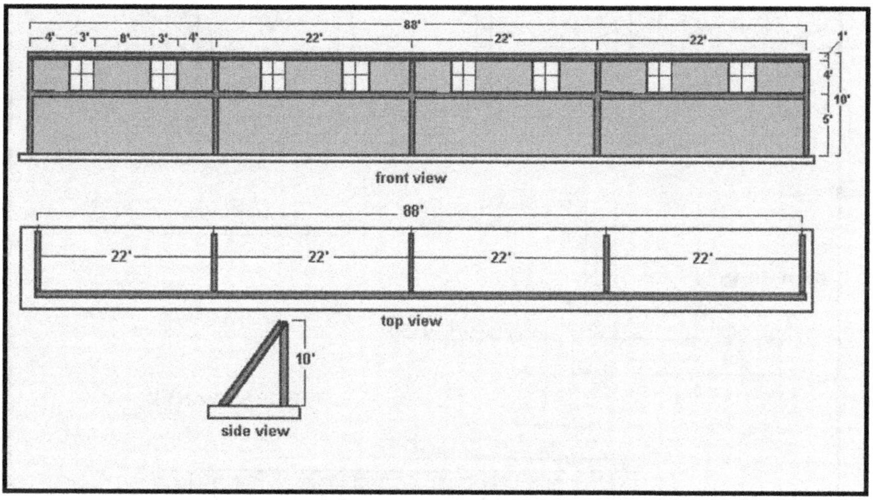

Figure 4-3. Station 2—Window breaching.

STATION 3

4-5. This station is a facade that looks like a wall of a building. It measures about 8 feet high by 28 feet wide. Precast panels will fit into each of the three sections. The panels measure 8 feet square and can be made of concrete, concrete blocks, stone, brick, wood or other material. The panels are designed for use with thermal and explosive breaching techniques (Figure 4-4).

TRAINING AUDIENCE

4-6. The facility provides Infantry, Engineer, MP, and Cavalry leaders with a facility to train and evaluate individual and leader breaching tasks. Unit training requirements determine the size of the unit and the tasks to be trained at this facility (Appendix B provides a unit training matrix). The individual tasks trained at the Breach Facility follow:

- 052-193-1003, *Prime Explosives Nonelectrically.*
- 071-440-0027, *Execute a Mechanical Breach.*
- 051-193-1055, *Construct a Nonelectric Initiating Assembly with Modernized Demolition Initiators (MDI).*
- 052-193-1312, *Construct Demolition Initiating Sets.*
- *Conduct a Thermal Breach* (Engineers only).

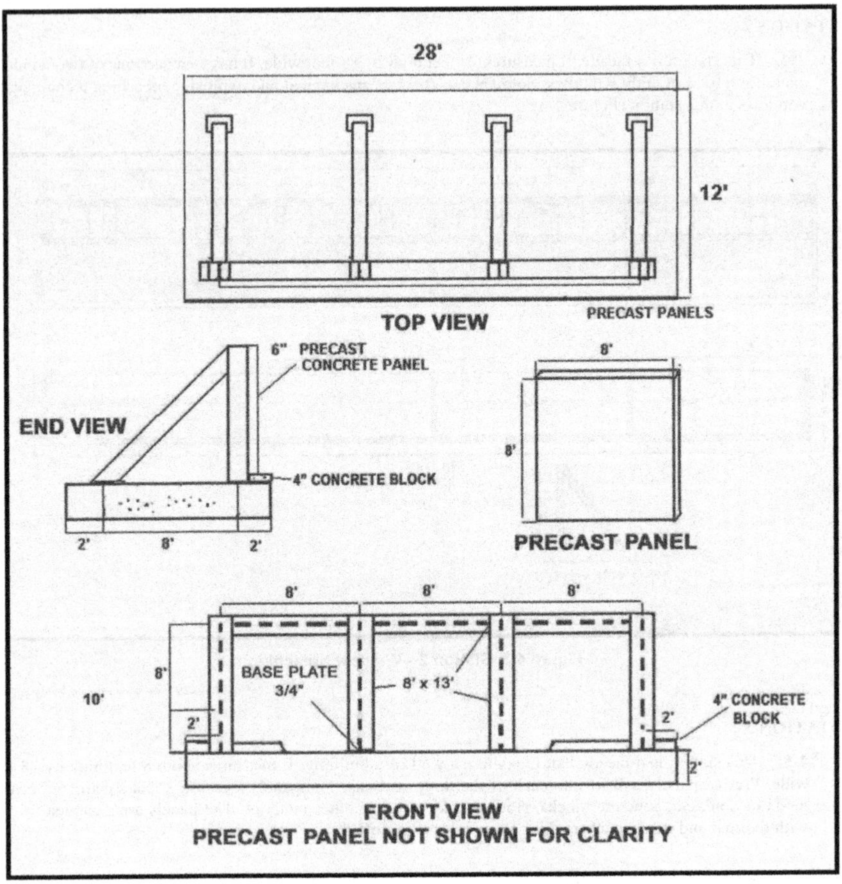

Figure 4-4. Station 3—Wall breaching.

TRAINING INFORMATION

4-7. Many types of organizations can train at this facility. Tactical operations in an urban environment may range from combat missions to stability and support actions. Commanders should note that this facility is an individual task and technique trainer. Training at the Breach Facility should be conducted prior to conducting collective training at the UAC, shoot house, and the CACTF. Trainers *must be certified* before conducting demolition training per local SOP or ACOM policy.

CONDUCT OF TRAINING

4-8. Trainers develop realistic time requirements for each task to be trained based on the *crawl-walk-run* concept and the size of the unit to be trained. The training event may be arranged in any order as long as it builds from simple to complex task and techniques. Table 4-1 contains an example sequence of events to assist units in developing their training plan. Leaders should do the following prior to conducting training. Before conducting ballistic breaching with automatic weapons, leaders should consult FM 3-06.11:

- Identify the unit's critical missions (see commander's training guidance) and select those tasks that support specific METL related to urban operations.
- Arrange the related tasks in the order of their importance to mission accomplishment.
- Assess the unit's current level of proficiency (strengths and weaknesses) and determine the degree of difficulty of each task. The training events should start from the less difficult tasks to the more difficult tasks. This instills confidence in the Soldiers conducting the training. Leaders should consult FM 3-06.11 for more information on explosive breach techniques.
- Identify the individual and leader tasks that are to be trained.
- Set up training conditions based on unit proficiency.

Table 4-1. Example platoon sequence of events.

Element	*Event Action*	*Estimated Time*
3d Sqd	Wall Breach: Explosive (crawl, walk, run)	3 hours
1st Sqd	Door Breach: Mechanical, ballistic, explosive (crawl, walk, run)	
2d Sqd	Window Breach: Mechanical, explosive (crawl, walk, run)	
2d Sqd	Wall Breach: Explosive (crawl, walk, run)	3 hours
3d Sqd	Door Breach: Mechanical, ballistic, explosive (crawl, walk, run)	
1st Sqd	Window Breach: Mechanical, explosive (crawl, walk, run)	
1st Sqd	Wall Breach: Explosive (crawl, walk, run)	3 hours NOTE: Walls, doors, and windows require replacement after each run/iteration to train to standards.
2d Sqd	Door Breach: Mechanical, ballistic, explosive (crawl, walk, run)	
3d Sqd	Window Breach: Mechanical, explosive (crawl, walk, run)	

TRAINING SUPPORT REQUIREMENTS

4-9. Table 4-2 shows minimum requirements based on three breach teams for each platoon-sized element. This facility is resource intensive, and Table 4-3 shows the total annual requirements per platoon-sized element.

Table 4-2. Consolidated minimum support requirements.

Ammunition/Equipment	*Crawl*	*Walk*	*Run*
Mechanical Breach: Prying , Striking, or Cutting Tool			
Door	(talk through)	1 per team (defeat one lock)	1 per team (defeat hinges)
Window	(talk through)	1 per team (defeat one lock)	1 per team (defeat hinges)
Ballistic Breach: Door			
Shotgun [1]	(talk through)	2 rds per weapon (defeat one lock)	5 rds per weapon (defeat hinges)
GREM (inert) [2]	(talk through)	1 per team – inert	1 per team - service
[1] If available to unit. [2] Grenade Rifle Entry Munition (GREM).			
Explosive Breach:			
Door: Det Cord (Flex Linear Charge) Nonelectric Firing System	(build)	(practice emplacement)	29 feet per team 1 per team
Wall: C4 (Satchel Charge) Nonelectric Firing System	(build)		4 blocks per team
		(practice emplacement)	23 feet per team 1 per team
Window: Det Cord (Flex Linear Charge) Nonelectric Firing System	(build)		29 feet per team 1 per team
		(practice emplacement)	
Thermal Breach Equipment Sappers (Only)	1	1	1

Table 4-3. Consolidated annual support requirements.

Ammunition/Equipment	*Total*
Doors	6
Windows	6
Wall Panels	6
Shotgun Rounds	42
GREM	6 inert/2 service
Det Cord (feet)	486
C4 (blocks)	24
Nonelectric Firing Systems	18

TRAINING AND EVALUATION

4-10. Individual training is the first step in the building-block approach and sets the basis for unit or collective training. Mastery of individual skills is an essential precondition for urban operations training and a good method for conducting individual training is through concurrent training. Concurrent training is designed to train groups of Soldiers on different tasks or techniques and these tasks or techniques may or may not be related. For example, a leader may subdivide the unit at the Breach Facility into breaching orders. One group of Soldiers may be conducting an explosive breach, while other groups are conducting mechanical and ballistic breaches. Ballistic breaching and explosive breaching does have a SDZ, an MSD, and ricochet considerations.

SAFETY CONSIDERATIONS

4-11. Safety requires effort on the part of everyone. It is not enough to teach safety rules; the rules must be practiced and enforced. Depending on the nature of the obstacle, the breacher may use mechanical, ballistic or explosive techniques. The following guidelines should be followed when training at the Breach Facility.

- Wear all appropriate personal protective equipment and follow appropriate guidelines:
  - -- Safety glasses.
  - -- Hearing protection.
  - -- Body armor.
  - -- Helmet.
  - -- Gloves.
  - -- Elbow and knee pads.
- Inspect equipment before use.
- Use appropriate tool.
- Use appropriate amount of explosive not to exceed four blocks of C4. Place the C4 on the wall only, not on the foundation, as this would cause irrepairable damage.
- Stay within the surface danger zone.
- Always use the minimum safe distance.
- Do not modify tool.
- Ensure proper use and maintenance of hydraulic and pneumatic tools; tools can burst under pressure when improperly used.

Chapter 5

Combined Arms Collective Training Facility

The home station CACTF is designed to conduct multiechelon, full-spectrum operations training up to battalion TF level. Example company and platoon STX lanes, and stability and support operations tasks, are provided to help units develop training plans. Leaders use training exercises to train, evaluate, and practice performance-oriented collective tasks. TEWT, STX, and FTX are the three types of exercises covered in this chapter. Each is designed to help the trainer develop urban operations training skills and increase unit proficiency in an urban environment. The urban operations collective training exercises were developed using the standard CACTF as described in this chapter and may be modified to fit available urban operations facilities.

PURPOSE

5-1. The leader or commander is provided with a CACTF to evaluate unit urban operations proficiency. The CACTF is intended to support blank fire, MILES/TES, SESAMS, STX, and FTX scenarios on a semiannual basis. The CACTF supports branch-specific lane training and combined arms training up to battalion level across the full spectrum of operations: offense, defense, stability, and support.

DESCRIPTION

5-2. The CACTF replicates an urban environment. The facility consists of 2.25 square kilometers of urban sprawl with 20 to 26 buildings, roads, alleys, parking areas, underground sewers, parks, athletic fields, and command and control building. The actual size and configuration of the CACTF depends on the local installation site requirements. The CACTF is designed to support heavy and light Infantry, Armor, artillery, and aviation positioning and maneuver. Table 5-1 shows the recommended types of buildings and features in the CACTF.

Table 5-1. CACTF characteristics.

20 to 26 Buildings (1.5 km by 1.5 km)	
Tunnel and sewer system Shantytown One 3-story building Three 2-story buildings Industrial area Electricity and potable water City dump	Props and furniture Targetry Audio and image capture with EDIT/REPLAY capability Control building with AAR facility Breachable walls

5-3. The buildings have one, two, or three stories (some with basements), sloped and flat roofs. CACTF building variations pose different tactical and technical training challenges. Figure 5-1 shows an example CACTF containing 24 buildings with the following characteristics:

- The hotel is a dominating structure typical of a central business district. The hotel has an elevator shaft, fire escapes, and a large first-floor lobby.
- The two-story office building contains a series of identical rooms.
- The school has a long, central corridor, large windows, and a single large room.
- The townhouse has multiple sections with a common attic.
- The service station accommodates combat vehicles in its service bay.

5-4. The buildings have other features that enhance the CACTF's training value.

- Constructed mouse-holes permit movement between rooms and floors.
- Loopholes in roofs and outside walls allow observation and fields of fire.
- Roof hatchways lead to the top floors of multiple story buildings.

5-5. Rooms vary in size, and in number of doorways and windows. These variations require Soldiers to identify rooms that provide suitable indoor fighting positions for TOW, Javelin, Dragon, and AT4 antitank weapon systems.

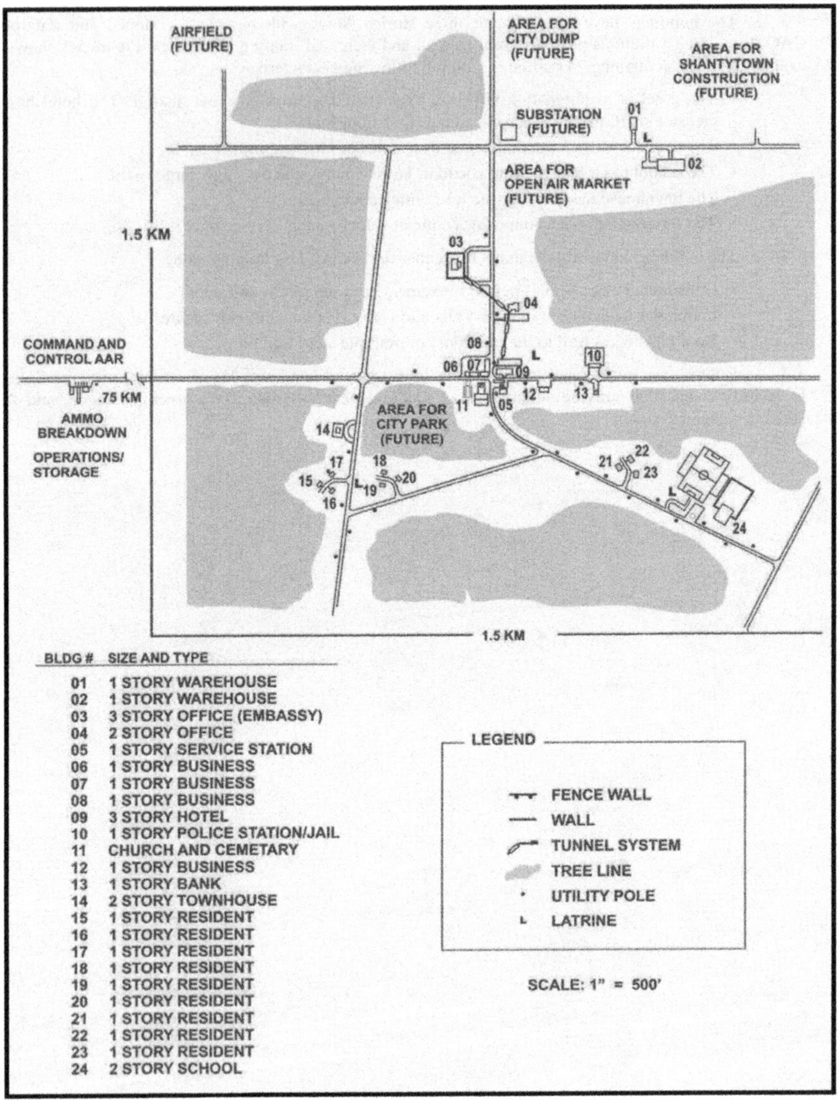

Figure 5-1. Example CACTF schematic.

5-6. Buildings are arranged in a realistic urban pattern. The CACTF represents an area consisting mainly of residential, commercial, public institutional, and light industrial buildings. Buildings and streets are arrayed with sufficient dispersion to allow up to nine platoon-sized STX sectors (Figure 5-2) or six company/team STX sectors (Figure 5-3).

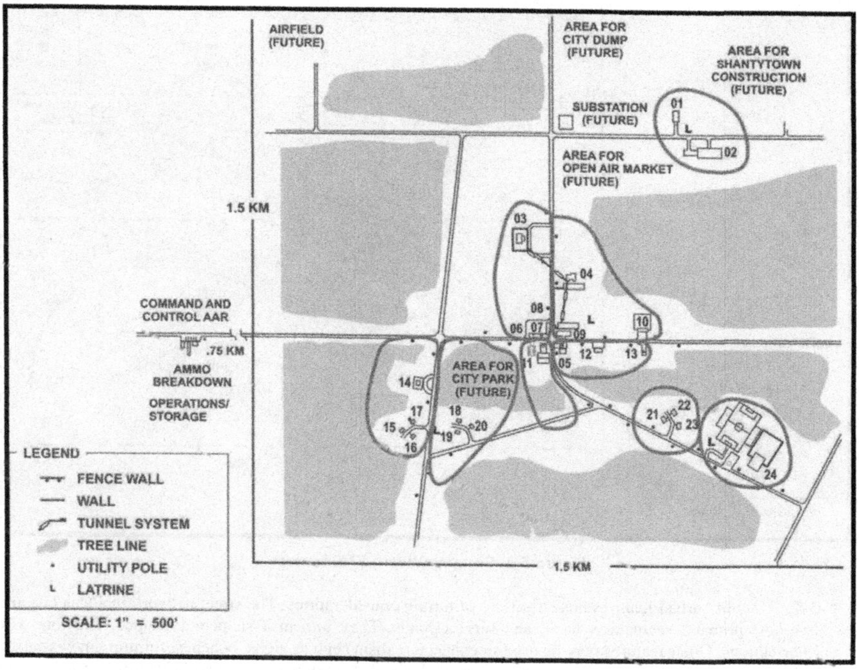

Figure 5-2. Platoon ODSS STX sectors.

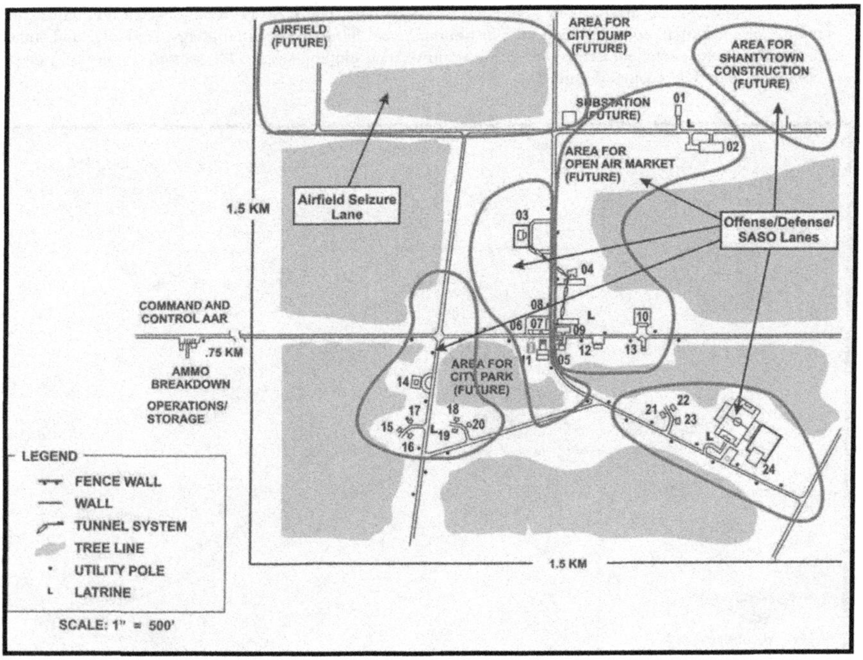

Figure 5-3. Company/team STX sectors.

5-7. Other urban features pose a variety of terrain considerations. The street network includes one and two-lane primary, secondary, local, and service roads. They join in T-shaped, L-shaped, and four-way intersections. Other features may be used to enhance realism such as signs, benches, furniture, newsstands, utility poles, street lights, vegetation, simulated rubble, live civilians on the battlefield, role players, and OPFOR. Open areas next to the CACTF can be used for maneuvering and staging areas. The specific details and layout of the CACTF may vary from one installation to another.

CAUTION

Trainers must ensure that fighting position reinforcement does not exceed the weight limits of the buildings in order to avoid structural damage and safety hazards to the Soldiers.

TACTICAL EXERCISE WITHOUT TROOPS

5-8. A terrain walk and a TEWT may be conducted in a local downtown area to prepare leaders and to take advantage of the available time in the CACTF. Terrain walks and TEWTs are conducted to teach leaders the tactical value of urban terrain. Appendix A provides training tips and additional information regarding the conduct of a TEWT.

COLLECTIVE TRAINING EXERCISES

5-9. The following exercises were developed IAW the guidance in FM 7-0, FM 7-1, FM 25-4, and TC 7-98-1. These publications include offensive and defensive scenarios for combat in urban areas, specifically, two platoon STXs and two company FTXs. Each can be conducted with blank ammunition; with MILES/TES; with 9-mm or 5.56-mm SESAMS for short-range precision target engagements (less than 50 meters); or force on force. Each was chosen to help the commander develop, sustain, and evaluate his unit's mission proficiency:

- Infantry Rifle Platoon Attack STX
- Mechanized Infantry Platoon Defend STX
- Infantry Rifle Company Attack FTX
- Mechanized Infantry Company Team Defend FTX

INFANTRY RIFLE PLATOON STX ATTACK

1. **Objective.**

This example STX trains collective, leader, and individual tasks for the platoon operation, Attack in a Built-Up Area.

2. **Interface.**

This STX is supported by two warrior drills, *React to Contact* and *Enter and Clear a Building*.

3. **Training.**

a. *Guidance*. FM 7-0 and FM 7-1 provide training guidance, tips, and enhancers for planning and executing STXs. The trainer should review the individual, leader, and collective tasks to be performed during the STX to determine the ones that require initial or refresher training.

b. *General Situation*.

(1) The exercise scenario follows: Contact with the enemy has been reestablished. Initial reports indicate he is at 65 to 70 percent strength, and that he has not received reinforcements. His defensive positions are in the CACTF and are not well established. He has indirect fire and CAS; he has already used chemicals, and will probably do so again. The platoon receives an order to prevent the enemy from establishing a heavily fortified defense in the CACTF. Such a defense could prevent friendly forces from continuing offensive operations.

(2) This exercise begins when the platoon leader receives the company FRAGO, and it ends after the platoon consolidates and reorganizes on the objective. An AAR should be held after completing the task, *React to Contact*, and at the completion of the exercise. If necessary, portions of the exercise should be repeated until the platoon performs them to standard. Figure 5-4 provides a graphic scenario of task performance for the STX *Attack*. Table 5-2 provides a recommended sequence for performance of T&EOs, and the time allotted for each portion of the STX.

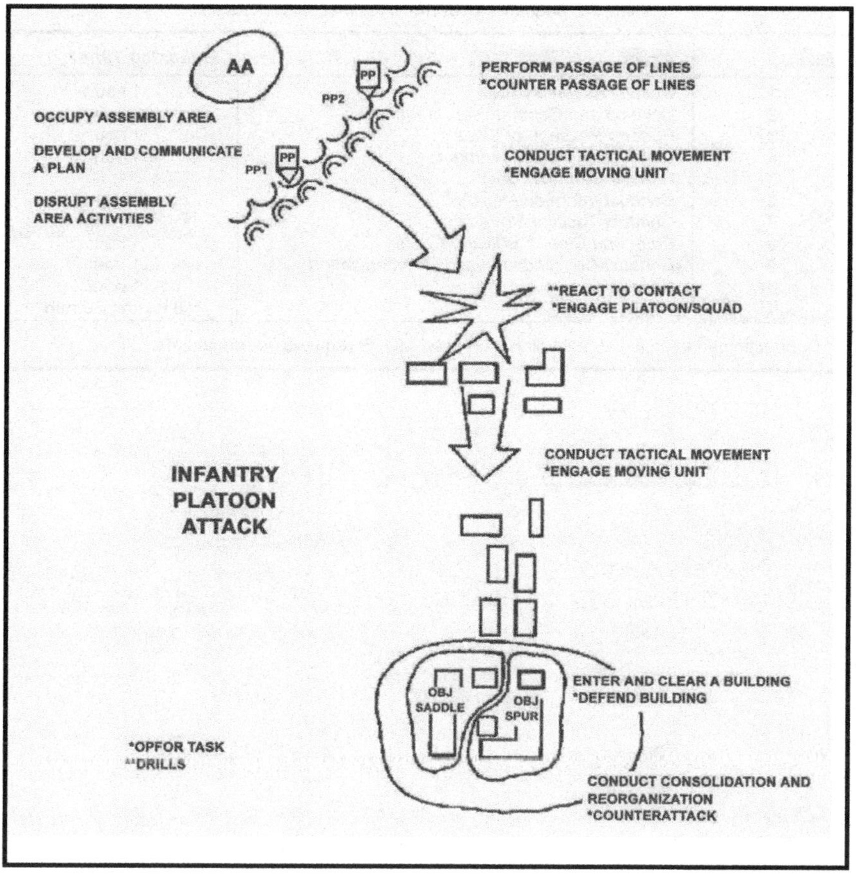

Figure 5-4. Attack STX scenario.

Table 5-2. Sequence of events and times for STX Attack.

Event	*Action*	*Estimated Time*
1	*Occupy Assembly Area*	1 hour
2	*Develop and Communicate a Plan*	1 hour
3	*Perform Passage of Lines*	1 hour
4	*Conduct Tactical Movement*	1 hour
5	*React to Contact (Drill)*	30 min
6	*Conduct Intermediate AAR*	30 min
7	*Conduct Tactical Movement*	30 min
8	*Enter and Clear a Building (Drill)*	1 hour
9	*Conduct Consolidation and Reorganization*	1 hour
10	*Conduct Final AAR*	1 hour
	*** Total Time**	**8 hours, 30 min**
* Conducting an exercise at night or in increased MOPP requires additional time.		

c. *Special Situation*. The platoon is part of a company in a secure assembly area when a FRAGO to attack is received (Figure 5-5).

1. **SITUATION.** The company is opposed by one platoon.

a. *Enemy Forces.* The enemy is at 65 to 70 percent strength. He is preparing a defense in CACTF (**MOUT site**) to prevent friendly forces from continuing to attack along the CACTF avenue of approach. He is expected to use artillery-delivered nonpersistent nerve agents in the defense of CACTF.

b. *Friendly Forces.* (Battalion designation) attacks (**date-time group**) to destroy enemy forces at OBJ COWBOY (**CACTF**) in order to pass follow-on forces, allowing friendly forces to continue the attack.

2. **MISSION.** (___) Company seizes OBJ HORSE (CACTF) vicinity (**grid**) NLT (**date-time group**) to prevent the enemy from establishing a heavily fortified defense.

3. **EXECUTION.**

a. *Concept of the Operation.* (See overlay.). The company moves in a wedge to Assault Position 1, assault CACTF to seize OBJ HORSE, consolidate, and reorganize.

(1) Maneuver. (**Evaluated**) Platoon will lead the company and be the main effort in the attack. They seize OBJ SADDLE vicinity (**grid**) to allow passage of friendly elements through CACTF. (___) Platoon moves on the company's left and seize OBJ SPUR vicinity (**grid**) to protect the left flank of the main effort. (___) Platoon moves on the company's right and establish a support-by-fire position located vicinity (**grid**).

(2) Fire support. fires are used to obscure movement to OBJ HORSE and to suppress the enemy during the assault. Priority of indirect fires is to (*evaluated*) Platoon, which is allocated one priority target (**120-mm mortar**).

Figure 5-5. Example FRAGO for STX Attack.

```
 b. Tasks to Maneuver Unit.
    (1)  (Evaluated) Platoon.
      (a)  Use Passage Point 1.
      (b)  Orient defense from TRP 2 to TRP 3.
      (c)  Be prepared to employ a wire-mined obstacle vicinity TRP 5
during consolidation.
    (2)  (___) Platoon.
      (a)  Use Passage Point 2.
      (b)  Orient defense from TRP 1 to TRP 2.
    (3)  (___) Platoon
      (a)  Use Passage Point 2.
      (b)  Orient defense from TRP 3 to TRP 4.
 c. Coordinating Instructions.
    (1)  Company RP is located (grid).
    (2)  Company linkup is (grid).
    (3)  MOPP1 is in effect (date and time).
    (4)  Selective rubbling is authorized. Destruction of an area
larger than one city block requires brigade commander approval.
```

Figure 5-5. Example FRAGO for STX Attack (continued).

4. **Support Requirements.**

a. *Minimum Trainers/Evaluators*: This exercise can be conducted with the company commander or platoon leader acting as the trainer or primary evaluator. At least one more controller or evaluator is required with the OPFOR. Other platoons being trained or evaluated will increase support personnel requirements.

b. *Vehicles/Communications*: All vehicles organic to the platoon should be included. The OPFOR should also have some combat vehicles.

c. *Opposing Force*:. The OPFOR should consist of at least a reinforced squad.

d. *Maneuver Area*: A training area at least 4 by 15 km is needed for cross-country movement and infiltration, and the CACTF is the ideal objective area. The terrain should offer multiple covered and concealed approaches to the objective.

5. **Evaluation.**

Table 5-3 lists the T&EOs from DTMS used to evaluate this STX.

Table 5-3. T&EOs used to evaluate STX Attack.

Task/Function	*Task Number*
Perform Passage of Lines	7-3-1125
Conduct Tactical Movement	7-2-1342
Occupy Assembly Area	7-3-1136
Develop and Communicate a Plan	7-3-1605
Conduct Consolidation and Reorganization (Company/Platoon)	7-2-5027

MECHANIZED INFANTRY PLATOON STX DEFEND

1. **Objective**. This example STX trains collective, leader, and individual tasks for the platoon operation, *Defend Urban Area/Building.*

2. **Interface.** This STX is supported by two warrior drills: *React to Contact* and *Enter and Clear a Building.*

3. **Training.**

a. *Guidance.* FM 7-0 and FM 7-1 provide training guidance, tips, and enhancers for planning and executing STXs. The trainer should review the individual, leader, and collective tasks to be performed during the STX to determine which require initial or refresher training.

b. *General Situation.*

(1) The exercise scenario is as follows: Latest INTSUMs indicate that enemy forces are expected to use a high-speed avenue of approach to counterattack. The approach is dominated by a small urban area containing strongly constructed buildings. The enemy has the capability for indirect fire, CAS, and chemicals. A defense in and around the town is necessary to deny the enemy the use of the avenue of approach. The platoon, as part of a larger force, is ordered to occupy and prepare defensive positions that overlook the avenue of approach. Indirect fire is available.

(2) Conduct of the exercise begins when the platoon receives an OPORD to move to establish a platoon defensive position. An AAR should be conducted after Soldiers enter a building and clear a room, and a final AAR should be conducted once all evaluation notes are compiled. If necessary, portions of the exercise should be repeated until the platoon's performance is satisfactory.

(3) When the platoon receives a warning order to prepare to move to establish a defense, the platoon is part of a company, which is part of a task force, and is located in an assembly area. Figure 5-6 provides a graphic scenario of task performance for the Defend STX. Table 5-4 provides a recommended sequence for performance of T&EOs and the time allotted for each portion of the STX.

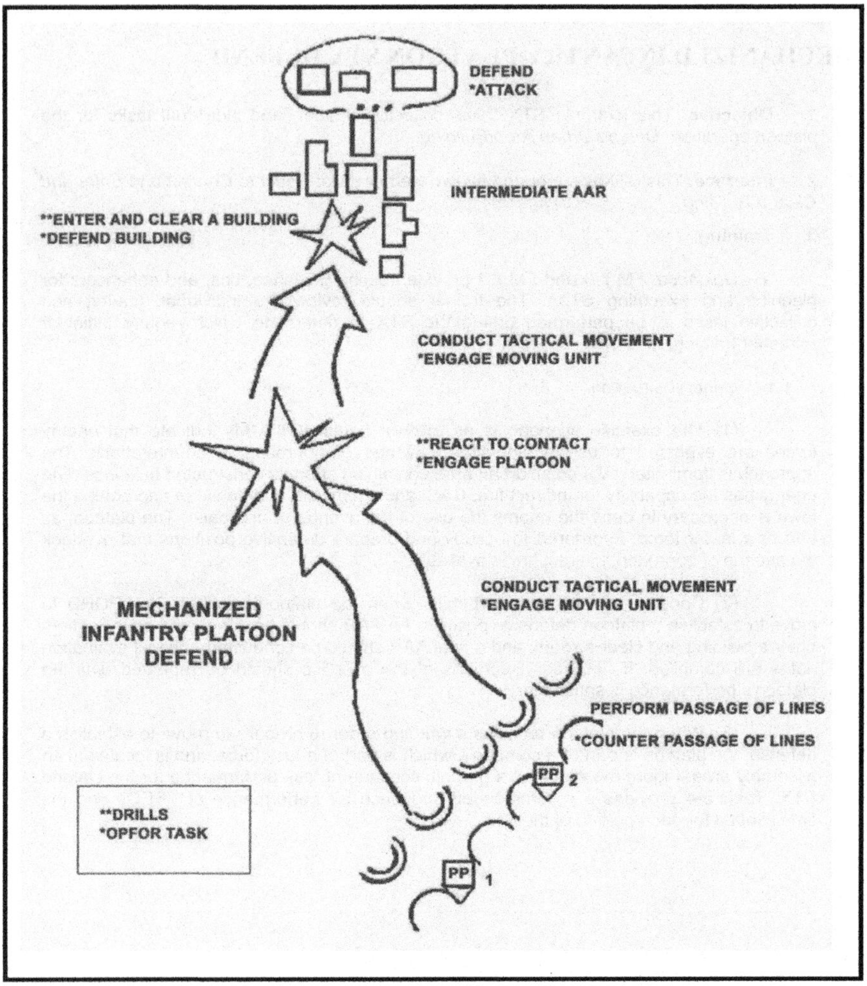

Figure 5-6. Defend STX scenario.

Table 5-4. Sequence of events and times for STX Defend.

Event	*Action*	*Estimated Time*
1	*Perform Passage of Lines*	1 hour
2	*Conduct Tactical Movement*	30 min
3	*React to Contact*	1 hours
4	*Conduct Tactical Movement*	30 min
5	*Enter and Clear a Building (Drill)*	1 hour
6	*Intermediate AAR*	30 min
7	*Defend MOUT/Building*	6 hours
8	*Conduct Final AAR*	1 hour
	***Total Time**	**11 hr 30 min**
*Conducting an exercise at night or in increased MOPP requires additional time.		

c. *Special Situation.* Thirty minutes after receiving the warning order, the platoon receives the company OPORD (Figure 5-7).

____
(Classification)

Copy no ___ of ___ copies
___ Battalion, ___ Infantry
Unit location or place of issue
Date-time group (with time zone)
Message reference number

OPERATION ORDER NO. ______
References: Map, series no. ___, sheet no. ___, edition ___
Time Zone Used Throughout the Order: ____
TASK ORGANIZATION:
___ **Platoon,** ___ **Platoon(-),** ___ **Platoon, Company Control**

Company HQ
1/1/C-52d Engr
1/2 ____,___
(Reserve)
Company Trains

1. **SITUATION.**

a. **Enemy Forces.**

(1) *Composition, Distribution, and Strength.* We are opposed by elements of the Liberation Guards Militia, Red Brigade. Several platoon-sized outposts are located within the battalion sector and consist of about 30 to 35 Soldiers per platoon. They are equipped with an assortment of Soviet small arms, including AK-47s, RPKs, RPGs, and various wheeled and tracked vehicles including some BMPs. Enemy outposts are at 60 to 70 percent strength and have indirect fire assets available.

(2) *Capabilities.* The enemy units are well entrenched in their defensive positions. The strongpoints have dedicated fire support assets and are capable of using nonpersistent nerve agents to prevent coordinated attacks. No enemy air threat exists but some heavy armored vehicles have been sighted in the area.

(3) *Probable Course of Action.* Intelligence indicates the enemy strongpoints will continue to fortify their positions until relief arrives within 24 hours in the form of a main enemy counterattack.

b. **Friendly Forces.** (__) TF defends in sector NLT (_______) along PL (_____) to deny enemy forces use of avenue of approach (_____). (__) TF must be prepared to continue movement forward of PL (_____), on order.

(1) Missions of units on left and right are (as required).

____
(Classification)

Figure 5-7. Example OPORD for STX Defend.

(Classification)

(1) (__) TF defends in sector NLT (______) from (grid) to (grid) to (grid) to contain any penetrations of the brigade sector.

(1) (___) Battery, (___) Battalion provides DS to 1st Brigade.

c. **Attachments and Detachments**. 1/1/C-52d Engineers attached.

2. MISSION. (___) Company blocks enemy forces NLT (date-time group) to prevent them from using high-speed avenues of approach through (name of MOUT site), and to prevent the enemy from counterattacking into the battalion sector.

3. EXECUTION.

a. **Concept of the Operation**. (See overlay.) Company prevents the enemy from using the high-speed avenues of approach through (MOUT site) by moving to the site and establishing a blocking position, which overwatches the platoon BPs covering the main avenues. We will block in depth by placing two platoons forward and one back, and maintaining a reserve to block any penetration. We must ensure our sectors of fire are placed to mass fire at the decisive point.

(1) *Maneuver*. (Evaluated) Platoon acts as the main effort, blocking from BP (__), vicinity (grid), to prevent enemy penetration of PL (____). (___) Platoon occupies BP (__), vicinity (grid), to block enemy movement into the main effort's left flank. (___) Platoon occupies BP (__) to prevent the enemy from making a coordinated attack on the main effort.

(2) *Fire Support*. The purpose of fires is to disrupt the enemy when his counterattack reaches PL (____). Priority of indirect fires is to (*evaluated*) Platoon, which is allocated two 155-mm priority targets, one of which may be used as an FPF.

(3) *Intelligence*. Priority of collection is to identify and locate enemy combat reconnaissance patrols and the enemy main effort.

(4) *Engineering*. During movement to the battle position, obstacles are used to ensure mobility. During blocking actions, the purpose of obstacles is to prevent the enemy from advancing on larger streets and to canalize him into more restrictive terrain. Priority of effort is to countermobility. Priority of support is to (*evaluated*) Platoon.

b. **Tasks to Maneuver Units**:

(1) *(Evaluated) Platoon*:

(a) Conduct passage of lines using PP 1.

(b) Lead the company movement after crossing the LD. Be prepared to react to enemy contact when (MOUT site) is reached.

(c) Enter and clear building(s) (number determined by commander).

(Classification)

Figure 5-7. Example OPORD for STX Defend (continued).

(Classification)

(d) Occupy BP (__) and orient on TRPs 2 and 3.
(e) Establish an OP vicinity (grid) to observe main avenue of approach.
(2) *(___) Platoon*:
(a) Conduct passage of lines using PP 2.
(b) Move second in order of movement.
(c) Enter and clear building(s) (number determined by commander).
(d) Occupy BP (__) and orient on TRP 1.
(3) *(___) Platoon*:
(a) Conduct passage of lines using PP 3.
(b) Move last in order of movement.
(c) Enter and clear building(s) (number determined by commander).
(d) Occupy BP (__) and orient on TRPs 4 and 5.
(e) Establish an OP vicinity (grid).

c. **Tasks to Subordinate Units**: Engineer squad travels with the company headquarters behind (*evaluated*) Platoon and supports the main effort's emplacement of obstacles.

d. **Coordinating Instructions.**
(1) Timings.
(a) Brief back ________.
(b) Company rehearsal ________.
(c) SP time ________.
(d) In position NLT (time) ________.
(2) Report all enemy contact.
(3) Report crossing all phase lines.
(4) MOPP status ____; enemy chemical attack is expected.
(5) Air defense status is YELLOW, WEAPONS HOLD.
(6) Leaders briefback the movement and defense plan to the commander at the company CP at (date-time group).
(7) Destruction of subterranean approaches requires brigade commander's approval.
(8) Selective rubbling is authorized. Destruction of more than one building requires commander's approval.
(9) Marking of rooms and buildings is IAW CO SOP.

(Classification)

Figure 5-7. Example OPORD for STX Defend (continued).

(Classification)

(10) Rules of Engagement follow:
(a) Minimize collateral damage to structures.
(b) Civilians have been evacuated. Evacuate to company CP any other civilians encountered.

4. SERVICE SUPPORT. IAW the battalion SOP.

a. **Miscellaneous**.
(1) Battalion trains located vicinity (grid).
(2) Company trains located vicinity (grid).

b. **Materiel and Services**.
(1) *Supply*.
(a) Class I. Ration Cycle M-M-M. Platoons must maintain a three-day supply in each BP.
(b) Class II. Submit requirements for special equipment ASAP. Coordinate with 1SG.
(c) Class III. Refuel all vehicles before (date-time group).
(d) Class IV. Coordinate with S-4 for delivery.
(e) Class V.

- Platoons cache one additional basic load in BP.
- Platoons submit request for additional demolitions NLT (date-time group).

(2) *Services*. Battalion decontamination site is located vicinity (grid).
(3) *Maintenance*.
(a) Priority to (*evaluated*) Platoon M2 BFVs.
(b) UMCP located vicinity (grid).

c. **Medical Evacuation and Hospitalization**.
(1) Battalion aid station located vicinity (grid).
(2) Company CCP located with company CP.

d. **Personnel**.
(1) No replacements for next 72 to 96 hours.
(2) Brigade mortuary affairs point located in BSA.
(3) EPW collection sites located at (grid) for battalion and (grid) for company.

(Classification)

Figure 5-7. Example OPORD for STX Defend (continued).

(Classification)

5. **COMMAND AND SIGNAL.**

a. **Command.** Initially, company CP will be traveling center of company wedge. Location in BP to be determined.

b. **Signal.**
(1) Current SOI in effect.
(2) Primary means of communication in BP is wire, FM, messenger, in that order.
(3) Radio listening silence is in effect until after crossing PL (____) or until contact is made.

(Classification)

Figure 5-7. Example OPORD for STX Defend (continued).

4. **Support Requirements.**

a. *Minimum Trainers/Evaluators*: This exercise is conducted by one company grade officer who is the trainer and primary evaluator.

b. *Vehicles/Communications*: All vehicles organic to the platoon should be included.

c. *Opposing Force*:. The OPFOR should be platoon strength.

d. *Maneuver Area*: A training area at least 2 by 10 km, with hasty fighting positions, is desirable for cross-country movement. The CACTF serves as the objective area. The terrain, ideally, should offer multiple covered and concealed routes. Using terrain that limits the leader to a geographic or predictable route prevents evaluation of the unit's ability to conduct a terrain analysis and select covered and concealed routes.

e. *Barrier Materials*. Sufficient Class IV (sandbags, lumber, and so on) must be made available to fortify positions, IAW Chapter 3, FM 3-06.11.

5. **Evaluation.**
Table 5-5 lists the T&EOs from DTMS used to evaluate this STX, and Table 5-6 lists the supporting tasks.

Table 5-5. T&EOs used to evaluate STX Defend.

Task/Function	*Task Number*
Defend MOUT/Building	7-3-1118
Perform Passage of Lines	7-3-1125
Conduct Tactical Movement (Co/Plt)	7-2-1342

INFANTRY RIFLE COMPANY FTX ATTACK

Table 5-6. Supporting tasks for company FTX.

STX C-II-1	*STX C-II-2*	*STX C-II-3*
Conduct an Infiltration (Co/Plt)	Perform Cordon and Search	Disarm Belligerents
Perform Linkup	Take Action on Contact	Handle Captured Insurgents or Belligerents
Execute an Assault (MOUT)	Conduct an Infiltration (Co/Plt)	Apprehend/Detain Noncombatants
Conduct Consolidation and Reorganization (Co/Plt)	Perform Linkup	
	Perform Passage of Lines	
	Occupy Assembly Area	

1. **Objective**. This FTX is designed to train the Infantry rifle company in conducting offensive operations. This exercise provides practice for the company commander, platoon leaders, squad leaders, and company headquarters personnel in planning, coordinating, and controlling combat operations.

2. **Interface**. This FTX is supported by STX C-II-1, C-II-2, and C-II-3 (Table 5-6). It is also supported by two warrior drills, *React to Contact* and *Break Contact*.

3. **Training**.

a. *Guidance*. FM 7-0 and FM 7-1 provide training guidance, tips, and enhancers for planning and executing STXs. The trainer should review the individual, leader, and collective tasks to be performed during the STX to determine which require initial or refresher training.

b. *General Situation.*

(1) The battalion is ordered to conduct offensive operations in support of the brigade mission. The company is ordered to infiltrate enemy lines and to destroy an insurgent operations base located in the CACTF. The company must be prepared to remain in the CACTF to conduct cordon and search missions.

(2) This exercise should be conducted under various conditions.

(3) When the company receives the battalion warning order the planning process for the attack mission begins. The exercise ends when the company completes a passage of lines, occupies an assembly area, and completes preparations to continue its mission. Figure 5-8 shows a graphic scenario of task performance for the FTX Attack. Table 5-7 provides a recommended sequence for performance of T&EOs and the time allotted for each portion of the STX, including AARs.

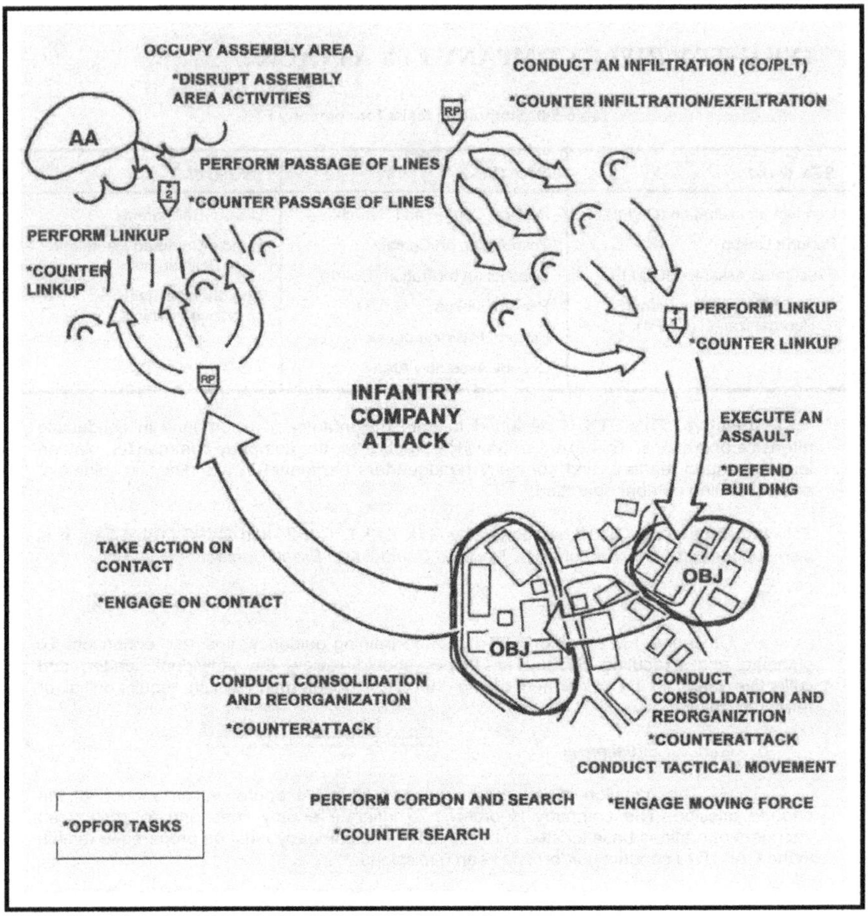

Figure 5-8. Attack FTX scenario.

Table 5-7. Sequence of events and times for FTX Attack.

Event	*Action*	*Estimated Time*
START STX C-II-1		
1	*Issue Company OPORD*	4 hours
2	*Conduct an Infiltration (Co/Plt)*	6 hours
3	*Perform Linkup*	3 hours
4	*Execute an Assault (MOUT) AAR*	3 hours
5	*Conduct Consolidation and Reorganization (Co/Plt)*	3 hours
ISSUE FRAGO, START STX C-II-2		
6	*Perform Cordon and Search*	6 hours
7	*Take Action on Contact / AAR*	3 hours
8	*Conduct an Infiltration (Co/Plt)*	6 hours
9	*Perform Linkup*	3 hours
10	*Perform Passage of Lines*	2 hours
11	*Occupy Assembly Area*	3 hours
START STX C-II-3		
12	*Disarm Belligerents*	3 hours
13	*Handle Captured Insurgents or Belligerents*	4 hours
14	*Apprehend/Detain Noncombatants*	4 hours
15	*Conduct Final AAR*	2 hours
	***Total Time**	**55 hours**

The time required to train an event varies based on the factors of METT-T and on the unit's training proficiency. Regardless, training is conducted to standards, not to time limitations. The following tasks are integrated and evaluated throughout the exercise:

- *Prepare for combat*
- *Maintain operations security*
- *Conduct tactical movement*
- *Perform consolidation*
- *Perform reorganization*
- *Perform stability and support operations*

* Time for intermediate AARs is excluded from this total; additional time is required if much of the exercise is conducted at night or during other limited visibility.

(4) The battalion commander issues an OPORD to the company commanders (Figure 5-9).

(Classification)

Copy no ___ of ___ copies
___ Battalion, ___ Infantry
Unit location or place of issue
Date-time group (with time zone)
Message reference number

OPERATION ORDER NO. _____
References: Map, series no. ___, sheet no. ___, edition ___
Time Zone Used Throughout the Order: ____

TASK ORGANIZATION:
___Company ___Company ___Company Battalion Control
Scout Platoon
81-mm Mortar Platoon
Antitank Platoon
1/C/58th ENGR (DS)
Battalion Trains

1. SITUATION.

a. **Enemy Forces**. Annex A (See intelligence overlay [prepared by brigade S-2].)

(1) *Disposition, Composition, and Strength*. The battalion is opposed by one company from the Dona Anna Militia Guards Battalion. This enemy company consists of 3 platoons with 35 to 40 Soldiers, equipped with M16s, AK-47s, and LMGs. The company headquarters is currently located in the town hall of CACTF. The company is at 90 percent strength and has indirect fire support assets available.

(2) *Capabilities*. The enemy possesses nonpersistent nerve agents and indirect fire assets capable of delivery. He also has some light armored vehicles in the area, but their location is undetermined. No enemy air threat exists. However, the enemy may be able to reinforce CACTF with an additional company, within two hours.

(3) *Most Probable Course of Action*. The enemy defends CACTF and continues using it as a base of operations and to consolidate weapons and material. His intent is to prevent the capture of CACTF and its support base, provided by its facilities and population.

(Classification)

Figure 5-9. Example battalion OPORD for FTX Attack.

(Classification)

b. **Friendly Forces.**

(1) (___) Brigade, (___) Division attacks at (date-time group) to seize the town of CACTF at (grid), to keep the enemy from using it as a base of operations. The intent is to secure the town and the road network in the area, to prevent the enemy from using it as a base of operations.

(2) (___) Battalion, (___) Infantry attacks to seize the road network at OBJ (____), vicinity (grid), at (date-time group) to prevent enemy forces from counterattacking along Route (____).

(3) (___) Battalion, (___) Infantry attacks at (date-time group), to seize the high ground at OBJ (____), vicinity (grid), and to establish defensive positions to block enemy routes in and out of CACTF.

2. MISSION. (___) Battalion, (___) Infantry attacks at (date-time group) to seize CACTF, vicinity (grid), NLT (date-time group), to prevent the enemy from using the town as an operations and logistics base.

3. EXECUTION.

a. **Concept of the Operation.** The battalion will accomplish its mission, which is to deny the enemy use of CACTF and its surrounding terrain and population, by seizing the town and controlling the key areas in the vicinity. We will conduct a feint in the west and then attack the decisive point within CACTF (OBJ [____]).

(1) *Maneuver.* (___) Company conducts a feint to deceive the enemy, then the main effort (*evaluated*) company seizes OBJ (____), to prevent the enemy from using CACTF as an operational and logistical base. (___) Company supports the main effort by seizing OBJ (____), in order to prevent the enemy from counterattacking the main effort. (___) Company seizes OBJ (____) to support the main effort's attack.

(2) *Fire Support.* Annex C. (See fire support.) Purpose of fires is to suppress the enemy and to obscure movement to CACTF. Priority of artillery fires is to (feinting) Company; on order, priority will shift to (supporting effort) Company. Priority of 81-mm mortar fires is to (*evaluated*) Company.

(3) *Engineering.* The purpose of the engineers is to breach obstacles during the attack, to help (*evaluated*) Company seize OBJ (____). Priority of support is to (*evaluated*) Company, (___) Company, and (___) Company in order. The battalion commander must authorize any booby traps, and point and hasty minefields.

(Classification)

Figure 5-9. Example battalion OPORD for FTX Attack (continued).

(Classification)

b. **Tasks to Maneuver Units.**

(1) *(Evaluated) Company.*

(a) Infiltrate using infiltration lane TANGO.

(b) Seize OBJ (____), vicinity (grid), NLT (date-time group).

(c) Be prepared to perform cordon and search of CACTF, after consolidation and reorganization.

(2) *(Supporting) Company.*

(a) Follow (main attack) Company and infiltrate on infiltration lane TANGO.

(b) Seize OBJ (____), vicinity (grid), NLT (date-time group).

(c) Coordinate the attack with (*evaluated*) Company, so that both units attack at the same time.

(d) Be prepared to perform cordon and search in the objective area after consolidation and reorganization.

(e) Be prepared to assume the mission of (*evaluated*) Company.

(3) *(Supporting) Company.*

(a) Infiltrate using infiltration lane ZEBRA.

(b) Conduct feint to OBJ (____), vicinity (grid), NLT (date-time group).

(c) Be prepared to establish defensive positions vicinity (grid) to prevent enemy movement to and from CACTF.

c. **Tasks to Subordinate Units.**

(1) *(Scout) Platoon.*

(a) During movement, determine location, strength, and disposition of enemy forces in CACTF.

(b) On order, screen the battalion flank from (grid) to (grid) to warn of the approach of enemy units from the southwest

(2) *Mortar Platoon.* Priority of mortar fire is to (*evaluated*) Company. Mortar platoon moves with (*evaluated*) Company, establishes a firing position vicinity (grid), and displaces on order.

(3) *Antitank Platoon.*

(a) Initially, over watch from (grid) to (grid) to secure the unit infiltration.

(b) On order, move vicinity (grid) to support (*evaluated*) Company's consolidation and reorganization.

(Classification)

Figure 5-9. Example battalion OPORD for FTX Attack (continued).

(Classification)

(4) *1/C/58 Engineers.*

(a) Move with (*evaluated*) Company to provide mobility support.

(b) Be prepared to reduce obstacles in CACTF during cordon and search mission.

d. **Coordinating Instructions.**

(1) Report when ready to execute an assault.

(2) Report any movement of enemy forces from the southwest.

(3) No rubbling is authorized, so minimize collateral damage to property.

(4) MOPP0 is in effect.

(5) Terrorist threat warning is red.

(6) Commanders brief back OPORD to battalion commander at battalion tactical CP at (date-time group).

(7) Rules of Engagement (ROE).

(a) Commanders take all steps necessary and appropriate for their units' protection.

(b) The minimum necessary force is used to control the situation.

(c) Leaders must take measures to minimize risk to civilians, without endangering the unit.

(d) Fire is returned directly to its source, not sprayed into a general area.

(e) Firing ceases when the threat is over.

(f) Anyone trying to surrender is allowed to do so.

(g) Civilians and property are treated with respect.

(h) WP can be used vicinity CACTF to aid in isolating the objectives. The requests for indirect fire, within the town, must be authorized by the battalion commander.

4. **SERVICE SUPPORT.** (See Annex D.)

5. **COMMAND AND SIGNAL.**

a. **Command.**

(1) The tactical CP moves with (*evaluated*) Company.

(2) The main CP is vicinity (grid) and displaces on order.

(3) The second in command remains in the battalion main CP.

(4) The rear CP is the alternate battalion main CP.

(Classification)

Figure 5-9. Example battalion OPORD for FTX Attack (continued).

(Classification)

b. **Signal.**
(1) Current SOI is in effect.
(2) Radio-listening silence is in effect for all elements except scouts, until prepared to attack.

ACKNOWLEDGE:
JONES
LTC
OFFICIAL:
SMITH
S-3
Annexes: A–Intelligence
B–Operations Overlay
C–Fire Support
D–Service Support

(Classification)

Figure 5-9. Example battalion OPORD for FTX Attack (continued).

(5) The battalion has destroyed the enemy operating in CACTF, and has seized the town. Two companies are defending against a limited counterattack. One company is in a blocking position. The (*evaluated*) company receives a FRAGO (Figure 5-10) to conduct a cordon and search operation. The purpose of the operation is to locate enemy personnel and equipment remaining in the CACTF.

Enemy personnel in squad-sized elements are reported by the civilian population to be hiding in the CACTF. Reports also indicate numerous weapons caches are being hidden by civilians collaborating with the enemy. Battalion has completed its mission and is preparing to conduct a cordon and search.

After completing this, the battalion exfiltrates to perform a passage of friendly lines. (Evaluated) Company moves to perform cordon and search vicinity OBJ (____), beginning (date-time group) to (date-time group), to locate and seize enemy personnel or equipment.

The (*evaluated*) company notifies the commander of the results of the search, and then notifies the commander when they (*evaluated* company) are ready to start movement. The company exfiltrates on order along Route Gold to perform linkup at (grid) and passage of lines at (grid). (See overlay for lanes.)

Figure 5-10. Example FRAGO for FTX Attack.

4. **Support Requirements.**

a. *Minimum Trainers/Evaluators*:. The company commander acts as the trainer-evaluator for the platoons. The battalion may direct the FTX internal evaluation along with a company evaluator and an FO evaluator. The platoon leaders and platoon sergeants are the primary trainers during the initial phases of the training. The company commander, the XO, the platoon leaders, and the platoon sergeants all act as trainers during the company phases of training. If the exercise is conducted by the battalion, either the battalion commander, XO, or S-3 acts as the trainer and evaluator, but additional battalion assets may be required to support the evaluation phase. The controllers or evaluators are positioned at critical locations throughout the exercise to observe the action. They must not interfere with the conduct of the operation, except to halt it when a safety violation occurs or to conduct an AAR.

b. *Vehicles/Communications*: Those organic to the company are used in the exercise. This includes one designated medical evacuation vehicle per platoon and at least one aidman on site.

c. *Opposing Force*:. The OPFOR ground force should be at least platoon-size. FTXs conducted by the company require internal resourcing; units should rotate to ensure total training. OPFOR tasks and standards are provided in each T&EO.

d. *Maneuver Area*: A training area 5 by 10 km which includes the CACTF is best for this exercise.

5. Evaluation.

Table 5-8 lists the T&EOs used to evaluate this STX.

Table 5-8. T&EOs used to evaluate STX Attack.

Task/Function	*Task Number*
Perform Cordon and Search	7-3-1105
Take Action on Contact	7-4-1107
Perform Passage of Lines	7 4-1125
Perform Linkup	7-4-1128
Conduct Tactical Movement	7-2-1342
Occupy Assembly Area	7-4-1136
Maintain Operations Security	7-4-1409
Prepare for Combat	7-4-1606
Conduct Consolidation and Reorganization (Co/Plt)	7-2-5027
Perform Cordon and Search	7-2-1105
Execute an Assault (MOUT)	7-2-1109
Perform Exfiltration	7-1-1146
Perform Infiltration	7-1-1147
Perform Reorganization	7-1-1608
Perform Consolidation	7-1-1609

MECHANIZED INFANTRY COMPANY OR TEAM FTX DEFEND

1. **Objective**. This FTX is designed to train the mechanized Infantry company or team in conducting defensive operations in urban terrain. It also provides platoons an opportunity to prepare for full-scale FTXs conducted and evaluated by the parent battalion. This exercise provides practice for the company commander, platoon leaders, squad leaders, and company headquarters personnel in planning, coordinating, and controlling combat operations.

2. **Interface**. This FTX is supported by the STX C-II-4, C-II-5, and C-II-6 tasks shown in Table 5-9. It is also supported by two warrior drills, *React to Contact* and *Break Contact*.

Table 5-9. Supporting tasks for company FTX.

STX C-II-4	*STX C-II-5*	*STX C-II-6*
Occupy Assembly Area	Perform Ambush	Operate a Checkpoint
Perform Tactical	Establish a Roadblock/Checkpoint	Interdict Smuggling Operations
Road March	Perform CBRN Operations	Defend Convoy
Perform Relief in Place	Conduct Tactical Movement	
Defend Built-Up Area/ Building	Perform Passage of Lines	
	Occupy Assembly Area	

3. **Training**.

a. *Guidance*. FM 7-0 and FM 7-1 provide training guidance, tips, and enhancers for planning and executing STXs. The trainer should review the individual, leader, and collective tasks to be performed during the STX to determine which require initial or refresher training.

b. *General Situation*.

(1) The battalion is ordered to conduct defensive operations in support of the brigade mission; the company is ordered to perform a relief operation and to prepare a defense in an urban area. The company remains in position for at least 36 hours before receiving a follow-on mission. Most civilians have been evacuated from the area, but a few remain. A terrorist threat exists as well as the possibility of enemy battalion-level offensive actions.

(2) This exercise should be conducted under various environmental conditions. The company also conducts operations under threat of CBRN attack.

c. *Special Situation.*

(1) The planning process for the defense mission begins when the company receives a warning order. The exercise ends when the company completes the passage of lines and occupies an assembly area. Figure 5-11 shows a graphic scenario of task performance for the FTX Defend.

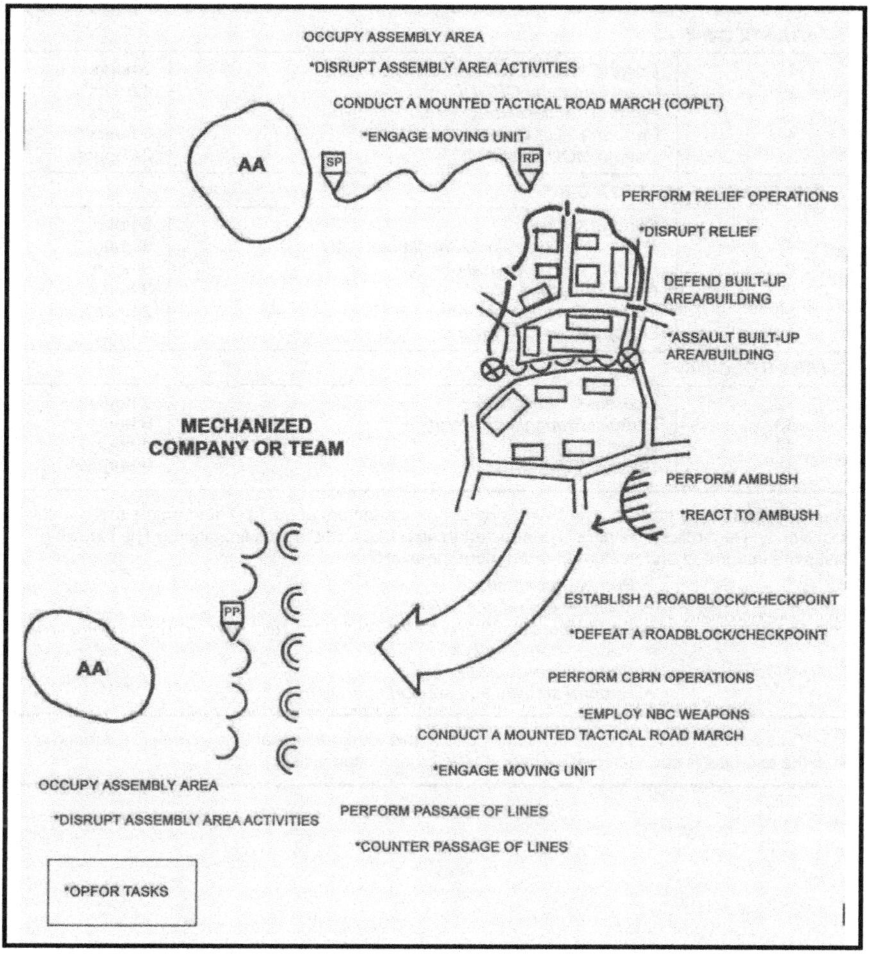

Figure 5-11. Graphic scenario of task performance for FTX Defend.

(2) Table 5-10 provides a recommended sequence for performance of T&EOs from DTMS and the time allotted for each portion of the STX, including AARs.

Table 5-10. Sequence of events and times for FTX Defend.

Event	Action	Estimated Time
START STX C-II-4		
1.	*Occupy Assembly Area*	2 hours
2.	*Issue Company OPORD*	4 hours
3.	*Perform Tactical Road March*	3 hours
4.	*Perform Relief Operations AAR*	6 hours
5.	*Defend MOUT/Building*	24 hours
ISSUE FRAGO, START STX C-II-5		
6.	*Perform Ambush*	6 hours
7.	*Establish a Roadblock/Checkpoint AAR*	6 hours
8.	*Perform CBRN Operations*	6 hours
9.	*Conduct Tactical Movement*	3 hours
10.	*Perform Passage of Lines*	2 hours
11.	*Occupy Assembly Area*	3 hours
START STX C-II-6		
12.	*Operate a Checkpoint*	2 hours
13.	*Interdict Smuggling Operations*	6 hours
14.	*Defend Convoy*	4 hours
15.	*Conduct Final AAR*	2 hours
The time required to train an event varies based on the factors of METT-T and on the unit's training proficiency. Regardless, training is conducted to standards, not to time limitations. The following tasks are integrated and evaluated throughout the exercise: ▪ *Prepare for combat* ▪ *Maintain operations security* ▪ *Conduct tactical movement* ▪ *Perform consolidation* ▪ *Perform reorganization* ▪ *Perform stability and support operations*		
* Time for intermediate AARs is not included in this total; additional time is required if much of the exercise is conducted at night or during other limited visibility.		

(3) The battalion commander issues the defense OPORD to the company (Figure 5-12).

```
                         ___________
                       (Classification)

                                          Copy no ___ of ___ copies
                                        ___ Battalion, ___ Infantry
                                   Unit location or place of issue
                                  Date-time group (with time zone)
                                          Message reference number
OPERATION ORDER NO. _____

Reference: Map, series no. ___, sheet no. ___, edition ___
Time Zone Used Throughout the Order: ____
TASK ORGANIZATION:
___Co ___Tm ___Tm (-)  Battalion Control
 1/__/__/AR 2/__/__/AR Scout Platoon
    120-mm Mortar Platoon
Team Tank E Co(-) 1/C/1-222 ADA (Stinger) (DS)
1/__/__/Inf  C/58 ENGR (DS)
2/__/__/Inf  1/1 (GSR) TM A/58 MI Bn
    ATK Force B (1/_/_/Inf)
    Battalion Trains
```

1. SITUATION.

a. **Enemy Forces**. Annex A (see intelligence overlay [prepared by brigade S-2]).

(1) *Composition, Disposition, Strength*. The company is opposed by the 1st Dragoons Regiment, 60th Guards Division. This regiment is at 85 percent strength and has received engineer and artillery reinforcements. The enemy is equipped with an assortment of former Soviet and Warsaw pact small arms, machine guns, and armored vehicles including T-62 tanks.

(2) *Capabilities*. The enemy has established defensive positions that follow a line from vicinity (grid) to vicinity (grid). Enemy armored reconnaissance units are operating in the more open terrain to the southwest of CACTF. The enemy has been observed moving forces in depth throughout the area in preparation for an offensive action.

(3) *Probable Course of Action*. The enemy's most probable course of action will be to begin offensive actions within the next 24 to 36 hours. Enemy encountered includes numerous reconnaissance elements trying to find the company's vulnerabilities and a main body trying to exploit any weakness.

____
(Classification)

Figure 5-12. Example battalion OPORD for mechanized company or team STX Defend.

(Classification)

b. **Friendly Forces.**

(1) (___) Brigade, (___) Division relieves (___) Brigade in place at (date-time group), to defend in sector from (grid) to (grid) to (grid) to (grid), in order to secure the corps counterattack axis of advance, (grid). The intent is to secure the road network around CACTF to prevent the enemy from impeding the movement of the (___) Brigade, (___) Division.

(2) TF (__) is the main effort in the defense and, NLT (date-time group), holds key terrain vicinity (grid) on our left, to prevent an enemy bypass of the brigade left flank.

(3) TF (__), on our right, defends along Route 360, vicinity (grid) to (grid) to (grid) to (grid), NLT (date-time group), to prevent an envelopment of the right flank. They conduct a supporting attack at (date-time group) to seize the high ground at OBJ OAK, vicinity (grid), to establish defensive positions that prevent the enemy armored reconnaissance units from interfering with the seizure of the pass at (grid) or the attack of the (___) Battalion, (___) Infantry, Mechanized.

2. **MISSION**. TF (____) defends NLT (date-time group), to retain the town of CACTF and its surrounding road network and allow brigade to prevent enemy interference with corps counterattack.

3. **EXECUTION.**

a. **Concept of the Operation**. Annex B (see operations overlay).

(1) Maneuver. Company (main effort) retains the main intersection (traffic loop) in CACTF, to prevent enemy movement through the sector. Team (*evaluated*) blocks movement west of 3d Street, to prevent the enemy from enveloping the main effort. The counterattack force is committed as needed. Team (__) blocks movement east of Oak Street, to canalize the enemy into EA (____). Team (__) defends from BP1, to prevent a coordinated attack against the main effort. E Company occupies BPs, to destroy enemy vehicles and protect the TF flanks.

(2) Fire Support. (See Annex C.) Fires are used to suppress the enemy and obscure our movement during the relief operation. Once positions are occupied, fires are used to suppress enemy overwatch positions, disrupt attempts to breach obstacles, and separate enemy Infantry from their supporting carriers. Priority of CAS and indirect fire is initially to Team (*evaluated*), during the relief operation, and then to Team (main effort). Teams (main effort) and (*evaluated*) each have two priority 155-mm targets. Each team may make one priority target an FPF.

(Classification)

Figure 5-12. Example battalion OPORD for mechanized company or team STX Defend (continued).

____
(Classification)

(3) Counter-Air Operations. Priority of protection is to the TF trains and main CP. Air defense weapons status is free; air defense warning status is yellow.

(4) Intelligence. Priority of collection is to identify and locate the C2 elements of the MRR and MRB, enemy tank battalions, and the enemy assault detachments' actual axes of advance.

(5) Engineering. The purpose of obstacles is to turn the enemy into restrictive terrain and engagement areas. Priority of support is to Team (main effort), Team (*evaluated*), Team (__), and Team (__), in that order.

b. **Tasks to Maneuver Units.**

(1) Team (Main Effort).

(a) Sector of defense includes the east side of Front Street, the south side of 1st Street, the west side of 3d Avenue area, and the north side of 3d Street.

(b) Team (main effort) relieves Team (__), TF (__).

(2) Team (Evaluated).

(a) Sector includes the east side of 3d Avenue, the south side of 1st Street, the west side of 6th Avenue, and the north side of 3d Street.

(b) Team (*evaluated*) relieves Team (__), TF (__).

(3) Team (__).

(a) Sector includes the east side of Oak Street, the south side of 1st Street, the east side of Front Street, and the north side of 3d Street.

(b) Team (__) relieves Team (__), TF (__).

(c) Team (__) coordinates directly with right flank team of TF (__), to ensure mutually supporting fires.

(4) Team (__)

(a) Occupy BP 1 vicinity (grid)

(b) Be prepared to occupy BP 2 (grid) to engage enemy in EA (____).

(5) Company E.

(a) Occupy BP 2-1 (grid) and move on order to BP 2-2 (grid).

(b) Provide antiarmor fires to destroy enemy tanks in EA (____).

(6) Counterattack B (Reserve).

(a) Occupy AA Black, vicinity (grid).

(b) Reconnoiter routes into TF (main effort) sector.

____
(Classification)

Figure 5-12. Example battalion OPORD for mechanized company or team STX Defend (continued).

(Classification)

(3) Counter-Air Operations. Priority of protection is to the TF trains and main CP. Air defense weapons status is free; air defense warning status is yellow.

(4) Intelligence. Priority of collection is to identify and locate the C2 elements of the MRR and MRB, enemy tank battalions, and the enemy assault detachments' actual axes of advance.

(5) Engineering. The purpose of obstacles is to turn the enemy into restrictive terrain and engagement areas. Priority of support is to Team (main effort), Team (*evaluated*), Team (__), and Team (__), in that order.

b. **Tasks to Maneuver Units.**

(1) Team (Main Effort).

(a) Sector of defense includes the east side of Front Street, the south side of 1st Street, the west side of 3d Avenue area, and the north side of 3d Street.

(b) Team (main effort) relieves Team (__), TF (__).

(2) Team (Evaluated).

(a) Sector includes the east side of 3d Avenue, the south side of 1st Street, the west side of 6th Avenue, and the north side of 3d Street.

(b) Team (*evaluated*) relieves Team (__), TF (__).

(3) Team (__).

(a) Sector includes the east side of Oak Street, the south side of 1st Street, the east side of Front Street, and the north side of 3d Street.

(b) Team (__) relieves Team (__), TF (__).

(c) Team (__) coordinates directly with right flank team of TF (__), to ensure mutually supporting fires.

(4) Team (__)

(a) Occupy BP 1 vicinity (grid)

(b) Be prepared to occupy BP 2 (grid) to engage enemy in EA (____).

(5) Company E.

(a) Occupy BP 2-1 (grid) and move on order to BP 2-2 (grid).

(b) Provide antiarmor fires to destroy enemy tanks in EA (____).

(6) Counterattack B (Reserve).

(a) Occupy AA Black, vicinity (grid).

(b) Reconnoiter routes into TF (main effort) sector.

(Classification)

Figure 5-12. Example battalion OPORD for mechanized company or team STX Defend (continued).

(Classification)

(c) Priority for planning:
- Battle position to reinforce main effort.
- Battle position to counterattack into Team (main effort) sector.
- Battle position to counterattack into Team (*evaluated*) sector.

(1) Scout Platoon.

(a) Determine location, strength, and disposition of enemy forces.

(b) Perform a screen of the TF's right flank from (grid) to (grid) beginning (date time group) to prevent enemy infiltration of CACTF.

c. **Tasks to Subordinate Units.**

(1) Mortar Platoon.

(a) Priority of fire to TF (main effort). Move with TF (main effort) and establish firing position vicinity (grid).

(b) Displace on order to (grid).

(2) 1/C/1 222 ADA. Coordinate locations with S-3 NLT (date time group).

(3) C/58 Engineers.

(a) Coordinate the engineer effort with the team commanders NLT (date time group).

(b) Provide an obstacle clearing team to Counterattack Force B, NLT (date time group).

(4) 1/1 (GSR) TM A/58 MI. Coordinate location of assets with the task force S-3 NLT (date time group).

d. **Coordinating Instructions.**

(1) Selective rubbling is authorized, but destruction of more than one building requires TF commander approval.

(2) Destruction of subterranean approaches requires TF commander approval.

(3) Commanders brief back OPORD to TF commander at the TF tactical CP, vicinity (grid), at (date time group).

(4) MOPP1 is in effect as of (date time group).

(5) Decontamination site established vicinity Miller Pond (grid).

(Classification)

Figure 5-12. Example battalion OPORD for mechanized company or team STX Defend (continued).

____
(Classification)

4. SERVICE SUPPORT. (See Annex D.)

a. **General.**
(1) Combat trains located vicinity (grid).
(2) Field trains located in the BSA, vicinity (grid).

b. **Materiel and Services.**
(1) Supply.
(a) Class I. Ration cycle C-A-C. Team establishes MRE cache sufficient for 48 hours in sector.
(b) Class III. Resupply available on request. Class III located in combat trains.
(c) Class IV. Materials are allocated as shown in Annex D.
(d) Class V. Each team establishes ammunition caches in sector, but should also have two basic loads of ammunition on hand.
(2) Transportation. The main streets are capable of two way heavy vehicle traffic (class 60). Secondary streets may be one way and of lighter construction (class 50).
(3) Maintenance.
(a) Priority by unit: Team (main effort), Team (*evaluated*), Team (__), Company E, Team (__).
(b) Priority by vehicle: M1A1, M2/M3, M113.

c. **Medical Evacuation and Hospitalization.**
(1) TF aid station located vicinity (grid).
(2) Civilian casualties should be transported by nonmilitary means to clinic, located vicinity (grid).
d. Personnel.
(1) No replacements for the next 96 hours.
(2) EPW collection point located vicinity (grid).
(3) Mortuary affairs point located in BSA, vicinity (grid).

e. **Civil-Military Cooperation.**
(1) Curfew in effect 1800 to 0800 daily.
(2) Majority of civilians have been evacuated. Those remaining have been notified of the risk involved.

____
(Classification)

Figure 5-12. Example battalion OPORD for mechanized company or team STX Defend (continued).

(Classification)

5. **COMMAND AND SIGNAL.**

a. **Command.**
(1) Brigade main CP located vicinity (grid).
(2) TF tactical CP located vicinity (grid).
(3) TF main CP located vicinity (grid).
(4) TF commander located at tactical CP.
(5) TF XO located at main CP.

b. **Signal.**
(1) SOI index 1 9 in effect.
(2) Primary means of communication is wire, FM, messenger in that order.
(3) Building and obstacle marking IAW TF SOP.

ACKNOWLEDGE:

WILLIAMS
LTC

OFFICIAL:

SMITH
S-3

Annexes: A-Intelligence
B-Operations Overlay
C-Fire Support
D-Service Support

(Classification)

Figure 5-12. Example battalion OPORD for mechanized company or team STX Defend (continued).

(3) The TF has retained the north half of the CACTF after the enemy attack. The enemy was defeated but has established squad-sized to platoon-sized strongpoints in the southern half of the CACTF. The enemy is expected to receive reinforcements within the next 12 to 24 hours. Theprevent enemy reinforcements from moving to the strongpoints to strengthen and resupply them, TF headquarters orders the team to conduct platoon-sized ambushes. They also order the team to prepare to establish a roadblock, to control refugee flow through the area of operations, before the team moves out of the area. Headquarters issues a FRAGO (Figure 5-13).

```
Numerous enemy strongpoints are located to your front, vicinity (grid), (grid), and (grid). We expect these strongpoints to receive reinforcements within the next 12 to 24 hours.

Your mission is to establish platoon ambushes on the likely avenues of approach to the strongpoints, NLT (date-time group), in order to prevent reinforcements or resupply to these strongpoints.

Notify the TF commander when you are prepared to move.

Coordinate with TF scouts for guides. On order, after return from ambush missions, establish roadblocks at (grid) and (grid). Be prepared to move within the next 36 to 48 hours to conduct further follow-on missions.
```

Figure 5-13. Example FRAGO for FTX Defend.

4. **Support Requirements.**

a. *Minimum Trainers/Evaluators*: This exercise should be conducted for practice with the company commander as the trainer-evaluator for his platoons. The battalion may direct the FTX (internal evaluation) using a company evaluator and an FO evaluator. The platoon leaders and platoon sergeants are the primary trainers during the initial phases of the training. The company commander, the XO, the platoon leaders, and the platoon sergeants act as trainers during the company phases of training. If the exercise is conducted by the battalion, either the battalion commander, the XO, or the S-3 acts as the trainer and evaluator, but may require additional assets from the battalion to support the evaluation phase. The controllers or evaluators must be positioned at critical locations throughout the exercise, to observe the actions. They must not interfere with the conduct of the operation, except to halt it when a safety violation occurs or to conduct an AAR.

b. *Vehicles/Communications*: Those organic to the company are used.

c. *Opposing Force*:. The OPFOR ground force should be at least platoon-size. FTXs conducted by the company require internal resourcing, and units should be rotated to ensure total training. T&EOs provide training tasks and standards for the OPFOR.

d. *Maneuver Area*: A training area at least 5 by 15 km is best for this exercise.

e. *Consolidated Support Requirements*: This exercise requires the items shown in Table 5-11.

f. *Barrier materials.* Sufficient Class IV must be made available to fortify positions, IAW FM 3-06.11.

Table 5-11. Ammunition and pyrotechnics support requirements for a six-day FTX.

Ammunition	*DODAC*	*Per Plt/Co/Bn FTX*
5.56-mm blank	1305A080	120 rounds per rifle
5.56-mm SESAMS		120 rounds per rifle
5.56-mm blank	1305A075	200 rounds per M249 LMG
5.56-mm SESAMS		200 rounds per M249 LMG
7.62mm	1305A111	200 rounds per BFV/M1
		400 rounds per M60/240B
ATWESS cartridge	1370L367	6 (Viper) (for LAW) 18 per Dragon 14 per TOW system 9 per OH-58D/AH-64
Pyrotechnic device	1370L602	6 per M1 2 per target
Hand grenade fuze (M228)	1330G878	150 per battalion
Stun grenade M84	1330GG09	150 per battalion
Smoke grenade, HC		150 per battalion
Smoke pot		6 per battalion

5. **Evaluation**.

Table 5-12 lists the DTMS T&EOs used to evaluate this STX.

Table 5-12. T&EOs used to evaluate STX Attack.

Task/Function	Task Number
Defend MOUT/Building	7-3/4-1118
Conduct a Dismounted Tactical Road March (Co/Plt)	7-2-1189
Conduct a Mounted Tactical Road March (Co/Plt)	7-2-1198
Perform Relief Operations	7-3/4-1124
Perform Passage of Lines	7-3/4-1125
Conduct Tactical Movement	7-2-1342
Occupy Assembly Area	7-3/4-1136
Establish a Roadblock/Checkpoint	7-3/4-1401
Perform CBRN Operations	7-3/4-1406
Maintain Operations Security	7-3/4-1409
Prepare for Combat	7-3/4-1606
Conduct Consolidation and Reorganization	7-2-5027
Perform Ambush	7-2-1106

TARGETRY

5-10. Table 5-13 lists the combination of precision and nonprecision targets to support the diversity of scenarios required to support combined arms and branch specific lane training. Precision targets, less than 50 meters, are used in and around buildings within the CACTF and reinforce precision marksmanship techniques described in FM 3-22.9 (Table 5-13).

Table 5-13. CACTF target requirements.

Event	Number Of Lanes	Human Targets		Vehicle Targets	Totals	
					HT	VT
Company/Platoon STX Offense	3/9	Precision	Nonprecision	9	45	9
		15	30			

CONSIDERATIONS FOR TRAINING WITH ARMOR

5-11. Built-up areas consist mainly of man-made features such as buildings, streets, and subterranean systems. These features of urban terrain create a variety of tactical problems and possibilities. In order to ensure that the tank platoon can operate effectively in the urban environment, the platoon observation and direct-fire plans must address the ground-level or surface fight (in streets and on the ground floor of buildings), the aboveground fight (intrasurface and supersurface), and the subterranean fight (Figure 5-14). Infantry can assist the tank platoon by—

- Locating targets for tanks to engage.
- Destroying antitank weapons.
- Assaulting enemy positions and clearing buildings with tank support.
- Protecting tanks from antitank fires.

Figure 5-14. Urban battlespace.

TRANSPORTING INFANTRY

5-12. At times, the tank platoon must transport Infantrymen (Figure 5-15). They only do so when not expecting contact. If the platoon is moving as part of a larger force, and if it is tasked to provide security for the move, then the lead section or element does not carry Infantry.

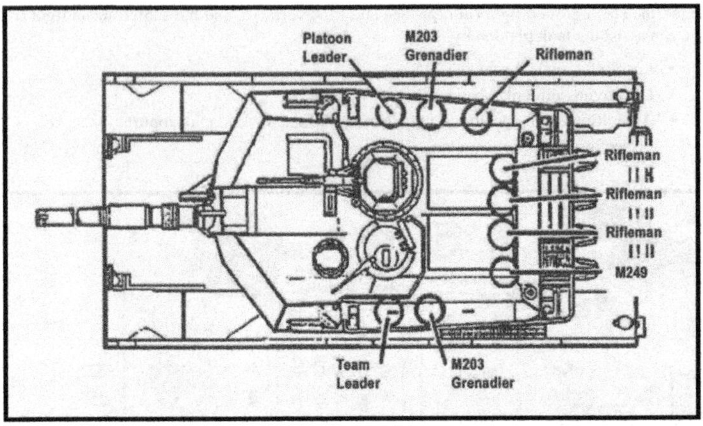

Figure 5-15. Example positions for Infantry riding on a tank.

5-13. Infantry and Armor leaders must observe the procedures, precautions, and considerations in Figure 5-16 when Infantrymen ride on tanks:

- Infantry should thoroughly practice mounting and dismounting procedures and actions on contact.
- Infantry must always alert the TC before mounting or dismounting. They must follow the commands of the TC.
- Infantry platoons should be broken down into squad-sized groups, similar to air assault chalks, with the Infantry platoon leader on the Armor platoon leader's vehicle and the Infantry PSG on the Armor PSG's vehicle.
- Platoon leaders, PSGs, and team leaders should position themselves near the TC's hatch, using the external phone (if available) to talk to the TC and relay signals to the unit.
- If possible, the lead vehicle should not carry Infantrymen. Riders restrict turret movement and are more likely to be injured or killed on initial contact.
- Whenever possible, Infantrymen should mount and dismount over the left front slope of the vehicle. This procedure ensures that the driver can see the Infantrymen and that the Infantrymen DO NOT pass in front of the coax machine gun. Infantrymen must ensure that they remain behind the vehicle's smoke grenade launchers. This will automatically keep them clear of all weapon systems.
- Infantrymen must always have three points of contact with the vehicle; they must watch for low-hanging objects such as tree branches.
- All Infantrymen should wear hearing protection.
- Infantrymen should not ride with anything more than their battle gear. Rucksacks and B-bags should be transported by other means.
- Infantrymen should scan in all directions. They may be able to spot a target the vehicle crew does not see.

Figure 5-16. Precautions for Infantrymen riding on a tank.

5-14. Passengers should be prepared to take the actions on contact shown in Figures 5-17 and Figure 5-18.

- Wait for the vehicle to stop.
- At the TC's command, dismount immediately (one fire team on each side). ***Do not*** move forward of the turret.
- Move at least 5 meters to the sides of the vehicle.
  Do not move behind or forward of the vehicle.
- ***Do not*** move in front of vehicles unless ordered to do so. Main gun discharge overpressure can inflict severe injury or death to forward dismounts.
- ***Do not*** dismount a vehicle unless ordered or given permission to do so.
- ***Do not*** dangle arms or legs, equipment, or anything else off the side of a vehicle; they could get caught in the tracks, causing death, injury, or damage to the equipment or vehicle.
- ***Do not*** place too many riders on the vehicle.
- ***Do not*** fall asleep when riding. The warm engine may induce drowsiness; a fall could be fatal.
- ***Do not*** smoke when mounted on a vehicle.
- ***Do not*** stand near a moving or turning vehicle at any time. Tanks have a deceptively short turning radius.

Figure 5-17. Actions on contact for Infantrymen riding on a tank.

DANGER

1. The overpressure from the tank 120-mm cannon can kill a dismounted Infantryman within a 90-degree arc extending from the muzzle of the gun tube out to 200 meters.
2. From 200 to 1,000 meters along the line of fire, on a frontage of about 400 meters dismounted Infantry must be aware of the danger from discarding sabot petals, which can kill or seriously injure Soldiers.

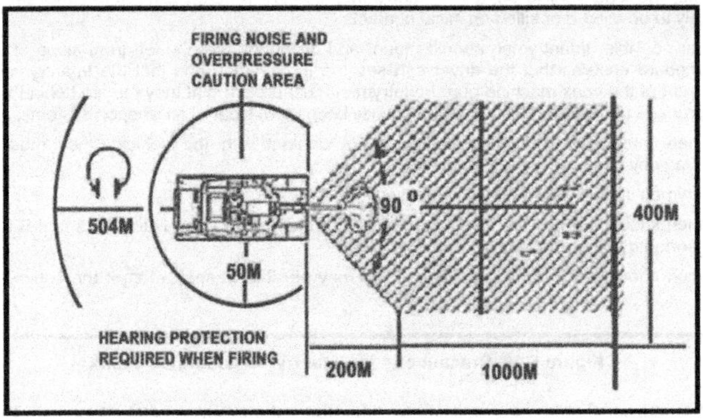

Figure 5-18. Danger areas around a tank when firing the 120-mm main gun.

ADDITIONAL CONSIDERATIONS

5-15. Additional considerations and preparations for transporting Infantrymen include the following:

- Armor--
  - Uses main gun fire to reduce obstacles or entrenched positions for the Infantry.
  - Takes directions from the Infantry ground commander (platoon leader/PSG/squad leader) to support their fire and maneuver.
  - Provides reconnaissance by fire for the Infantry.
  - Should know and understand how the Infantry clears buildings, how they mark cleared buildings, casualty evacuation plan, signal methods, engagement criteria for tank main gun, front line trace reporting, ground communication from the tank with the dismounted Infantry.
  - Uses its night vision capability to augment and supplement the Infantry's night vision capability.
- Infantry--.
  - Provides real time information for the tank crewmen to help them overcome tank noise and the lack of ground situational understanding.
  - Provides reconnaissance and fire direction of enemy positions for main gun attack.

- Considerations for dismounted tank security include--
  - -- Tank crewman should rehearse the mounting and dismounting of Infantry from their vehicle, briefing the Infantrymen on safety procedures for the vehicle and weapon systems.
  - -- Tank commanders need to rehearse communicating with dismounted Soldiers by way of TA-1 and DR-8 in the bustle rack.
- Vehicle preparation for combat in urban terrain should cover--
  - -- Keep at least one ballistic shield to the "Dog House" closed. Most engagements will be under boresight range, and the battlesight technique will suffice.
  - -- Place sandbags around antenna connections and electrical wiring on the turret top.
  - -- Place extra coax ammunition inside the turret.
  - -- Remove any highly flammable products from the outside of the vehicle and from the sponson boxes.

VEHICLES, WEAPONS, AND MUNITIONS

5-16. Numerous factors related to vehicles and their organic weapons and munitions affect the tank platoon's urban operation planning and execution, including the following:

- The preferred main gun rounds in the urban environment are HEAT, MPAT (ground mode), and MPAT-OR. These all perform much better than sabot rounds against bunkers and buildings.
- HEAT ammunition will open a larger hole in reinforced concrete or masonry structures than MPAT or MPAT-OR. Both MPAT and MPAT-OR, however, offer greater incapacitation capability inside the structure.
- HEAT ammunition arms approximately 60 feet from the gun muzzle. It loses most of its effectiveness against urban targets at ranges of less than 60 feet.
- MPAT and MPAT-OR rounds arm approximately 100 feet from the muzzle of the gun. Because of the shape and metal components of the projectiles, however, this ammunition remains effective at ranges of less than 100 feet.
- Sabot petals, including those on MPAT and MPAT-OR, endanger accompanying Infantry elements. They create a hazard area extending 70 meters on either side of the gun-target line out to a range of 1 kilometer.
- The tank's main gun can depress only to -10 degrees and can elevate only to +20 degrees. This creates considerable dead space for the crew at the close ranges that are typical in the urban environment.
- The external M2 HB machine gun can deliver a heavy volume of suppressive fire and penetrate light construction, buildings and most barricades. The M2 HB MG can elevate to +36 degrees; however, the TC must be unbuttoned to fire the M2 on the M1A2 or M1A2 SEP.
- The M240 coax machine gun can effectively deliver suppressive fires against enemy personnel and against enemy positions that are behind light cover.
- The loader's M240 machine gun can effectively deliver suppressive fire against enemy personnel and against enemy positions that are behind light cover; however, the loader must be unbuttoned to operate it. This weapon may be dismounted and used in a ground role if units are equipped with the M240 dismount kit.

- When buttoned up, the tank crew has limited visibility to the sides and rear and no visibility to the top. Figures 5-19 and 5-20 show the dead spaces associated with tank operations in an urban environment.
- FM 3-20.12 explains special uses for tank-mounted machine guns in the urban environment.

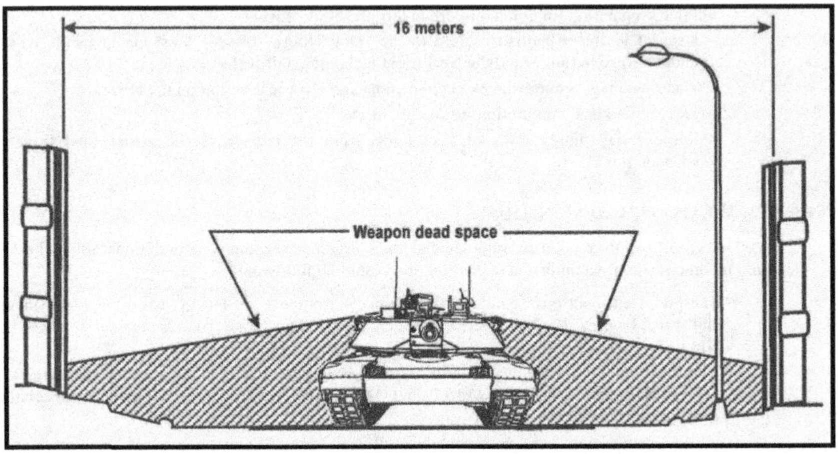

Figure 5-19. Tank weapon dead space at street level.

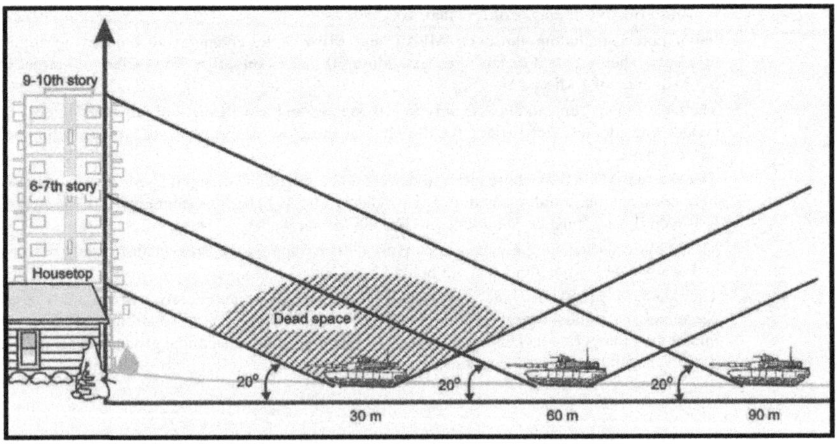

Figure 5-20. Tank main gun and coax dead space above street level.

TANK PLATOON COMMAND AND CONTROL IN URBAN OPERATIONS

5-17. Communications problems, fire control, proximity and visibility, personnel factors, ROE, ROI, civilians, and slow-pace urban operations all affect the tank platoon's urban operations planning and execution:

Communications Problems

5-18. The low-level task organization that may take place during urban operations requires elements to establish additional communications links, which can be disrupted by buildings and other urban terrain features.

Fire Control

5-19. Extensive direct fire planning and restrictive fire control measures are an absolute requirement in urban operations.

Proximity and Visibility

5-20. Friendly elements often must operate in confined and restrictive areas during urban operations, and they may not be able to see other nearby friendly forces. These factors significantly increase the danger of fratricide.

Personnel Factors

5-21. Urban operations impose significant, and often extreme, physical and psychological demands on Soldiers and leaders.

ROE, ROI, and Civilian Considerations

5-22. The ROE or ROI may restrict the use of certain weapon systems and TTP. As an integral part of urban operations, noncombatants create special operational problems. Thedeal with these concerns, units operating in urban terrain must know how to effectively employ linguists and counterintelligence and civil affairs teams.

Slow Pace Urban Operations

5-23. This operation usually prevents the tank platoon from taking full advantage of the speed and mobility of its tanks.

Appendix A
Training Tips

The effectiveness of urban operations training depends on the imagination and thoroughness applied to a training plan. All trainers, from training managers to junior leaders, should prepare their units to fulfill their urban operations mission. They can employ procedures that are based on training doctrine and proven techniques. This appendix discusses several techniques for conducting urban operations training. Trainers make the choice based on the training needs and available resources.

CONCURRENT TRAINING

A-1. Concurrent training allows groups of Soldiers to train on different tasks at the same time. This training requires a plan for rotating Soldiers between the stations that teach different tasks. This procedure is referred to as round-robin training. Concurrent training is an efficient training technique since it makes the best use of training time and facilities. Concurrent training in the urban operations training facilities works best with performance-oriented training. It also requires detailed planning, control, coordination, and assistant trainers. Each station should conduct training that is relevant to, and supports the training the unit is conducting.

URBAN ASSAULT COURSE AND BREACH FACILITY

A-2. These facilities are well suited for conducting concurrent training. After mastering the tasks at one station, a group of Soldiers rotates to the next station in the sequence. Theavoid bottlenecks at any of the stations, trainers plan instructional blocks that last about the same length of time. Concurrent training is often conducted with dedicated instructors at each station. The *advantage* of this technique is that it ensures uniformity of performance standards and saves on trainer preparation time. The *disadvantage* is that the Soldier is not always trained by his leader. Therefore, when feasible, small-unit leaders should rotate with their Soldiers from station-to-station so they can take part in the training. This technique demands prior planning, preparation, and instructors that can teach.

URBAN ASSAULT COURSE AND COMBINED ARMS COLLECTIVE TRAINING FACILITY

A-3. These facilities are well suited for conducting concurrent exercises. Concurrent exercises are variants of concurrent training. In a concurrent exercise, elements of the same unit receive different missions. These missions correspond to each element's training objective for that exercise. This technique can be used during FTXs, STXs, or CFXs. For example, one unit attacks; a second unit supports by fire; a third unit acts as a reserve. Also, elements of the same unit can challenge each other. For example, if a trainer has been training his unit in urban offense and defense operations, he can create a scenario in which one force is required to attack the other. This technique allows the unit to train with an OPFOR without an extra investment of resources. Leaders must remember to carefully plan the force ratios between the OPFOR and counterattacking forces.

DEMONSTRATIONS

A-4. Trainers must demonstrate the correct way to perform the tasks that the Soldiers will be required to perform. A demonstration is more effective when the trainer uses hand-on techniques. A live demonstration is an effective teaching method since it holds Soldiers' interest and they are able to learn with their eyes. The UAC and the Breach Facility are ideal facilities for teaching individual and collective tasks using demonstrations. The trainer must ensure that each Soldier. Performs the task being taught to guarantee that the soldier has learned. Key points should be explained to the satisfaction of each Soldier before progressing to the next step. The trainers can use video cameras to film the action occurring inside the buildings to show the Soldiers the correct way to perform a task or use the same action to explain a better technique to accomplish the same task. This technique is very effective because the Soldiers can watch the demonstrations on a remote monitor, and trainers can replay the videotape as often as required to reinforce his training objective.

EXERCISES

A-5. Training exercises are used to develop tactical SOPs, teamwork among the participants, and Soldier confidence in his leaders and equipment. Table A-1 matches training exercise types to appropriate training facilities.

Table A-1. Training exercise matrix.

EXERCISE	*CACTF*	*SHOOT HOUSE*	*SIMULATIONS*	*LOCAL TOWN*
MAPEX	X		X	X
TEWT	X			X
CPX	X		X	X
CFX	X			X
BSX			X	
LFX		X		
STX	X	X		X
FTX	X			X

MAP EXERCISE

A-6. A MAPEX allows participants to plan and execute urban operations using maps, overlays, sand tables, photographs, or models. MAPEXs develop teamwork and skills in terrain analysis, planning, coordination, order writing, and employment of urban operation tactics and are more suited for battle staff and leaders.

- Using a model of the CACTF is an excellent training aid for instructing Soldiers and leaders. Gather leaders around the map or model and have them devise a tactical plan. Use figures or markers to represent friendly units and the OPFOR. Create tactical situations that challenge participants to react as they would in real combat.
- Incorporate direct fire, casualties, CBRN, logistics, and civilians into the exercise. Ensure the ROE for each exercise is clear to participants.
- Have plenty of detailed maps available for use in the MAPEX and in the planning and execution of subsequent CFX operations.

TACTICAL EXERCISE WITHOUT TROOPS

A-7. A TEWT focuses on leader and staff tasks. The senior trainer selects the terrain for the urban TEWT. Either the CACTF or a real village, town, or city is appropriate. The procedure for conducting a TEWT is as follows:

- Create a scenario and enemy situation, and assign missions.
- Allow the leaders to devise their reconnaissance plans and implement them on the terrain.
- Require the leader to brief subordinates while walking the terrain.
- Ensure their plans include the following tasks:
  -- Analyze the terrain.
  -- Employ units according to the terrain analysis.
  -- Emplace weapons.
  -- Devise a unit plan.
- Conduct an AAR.

COMMAND POST EXERCISE AND BATTLE SIMULATION EXERCISE

A-8. CPXs and BSXs allow leaders and battle staffs to perform their tasks in a unit CP, where they receive information as reports requiring certain reactions. This type of training can be achieved through the constructive simulations training portion of the urban operations training strategy using JANUS or JCATS. A company CP is usually associated with a battalion level or higher CPX, but a company can conduct its own CPX. The headquarters section can set up a CP in a CACTF building, displace to another building, and repeat the exercise until they meet the required standards. The section can practice hiding wire on pavement, stringing it between buildings, and running it through sewers. The CP can practice receiving reports, posting situation maps, and sending reports forward to higher headquarters. The CPX/BSXs are most effective in achieving the following general training objectives:

-- Preparing plans.
-- Issuing orders.
-- Building teams.
-- Reconnoitering, selecting, and tactically occupying CP locations.
-- Establishing and using communications.
-- Preparing and sending reports.
-- Coordinating between subordinate elements.

COMMAND FIELD EXERCISE

A-9. The CFX is used to train commanders, staffs, and junior leaders. The CACTF or a local town can accommodate this type of exercise. However, the unit commander must go through the correct procedures when conducting training off federal property. The CFX employs all of a unit's command and control assets, with a number of subordinate elements reduced according to a set scale. For example, one Soldier might represent an entire squad, one vehicle a platoon, or one gun an entire battery.

Preliminary Training

A-10. Preliminary training, such as CPXs, TEWTs, MAPEXs, and BSXs, should precede the CFX. Units conduct operations, and controllers resolve battle outcomes. The exercise ends with an AAR. Fewer controllers are needed in a CFX than in an FTX. However, if the number of unit functions increases, the number of controllers must increase also.

Planning Procedures

A-11. Combat in urban areas entails centralized planning with decentralized execution.

Centralized

A-12. Centralized planning can be practiced in a CFX; however, it can be practiced more efficiently in a CPX, TEWT, MAPEX, or BSX.

Decentralized

A-13. Decentralized execution means that platoon, squad, and team leaders lead the battle in an urban area. However, because element participation is scaled down in a CFX, trainers usually find STXs, FTXs, and TEWTs more useful.

LIVE-FIRE EXERCISE

A-14. Chapter 3 and local range regulations provide specific guidance for conducting LFXs at the shoot house. Live-fire training allows Soldiers and leaders to maneuver while firing service ammunition or SRTA. An LFX is most productive when preceded by non-live-fire practice (dry fire or dry run). However, dry fire should *always* precede live fire, with either TES or SESAM. An AAR should follow each repetition, and units should only proceed to the next level of difficulty after mastering the current step. The training in an LFX occurs in the following sequence, which can be shortened if time and ammunition dictate.

1. Dry fire to practice unit drills and individual tasks.
2. Conduct an AAR.
3. Practice drills, tasks a second time with limited use of ammunition to show the complex coordination of fire and maneuver required.
4. Conduct an AAR.
5. Practice exercise a third time with full use of ammunition to enhance realism and build confidence.
6. Conduct an AAR.

SITUATIONAL TRAINING EXERCISE

A-15. STXs are mission-oriented, limited exercises used to train a unit to perform a single collective task or a series of related tasks and drills. With their distinct start and stop points, STXs represent a segment of battle. An STX is more flexible and complex than a single T&EO or drill. In fact, an STX typically includes drills, leader tasks, and separate individual tasks, and involves a unit's "slice" of external combat assets. An STX often precedes an FTX or CFX, and can be conducted in the CACTF by creating a brief situation and mission. For example, the MP company commanders might want to teach his Soldiers about straggler control. They can string the tasks together into small scenarios such as STXs, describe a tactical situation that produces stragglers from nearby units, use other Soldiers to represent the stragglers, and train Soldiers to process stragglers at straggler collection and transport points.

FIELD TRAINING EXERCISE

A-16. FTXs allow units to execute collective tasks under conditions that closely resemble actual urban terrain. FTXs are suitable for the CACTF.

- An FTX requires significant time, resources, and planning. The time is well spent if units master their collective tasks. An FTX is an excellent way to conduct the *run* stage of collective training. Additionally, OPFOR and MILES greatly enhance the training value of the FTX.
- A unit's urban operations FTX training plan addresses ROE, maneuver restrictions, and the controller plan. Soldiers must know the exercise's training objectives before they begin, so they can focus on the correct tasks. Every participant, regardless of position or rank, must fully understand all safety requirements.

- To make the most of their training time in the field, leaders must plan in great detail. The concept of multiechelon training demands that they consider what training can occur at each level in each event.

EVALUATORS

A-17. Soldiers from the training unit or from another unit can evaluate urban operations training, especially during the *run* stage of training in an STX, FTX, or CFX. A dedicated evaluator provides objective and thorough comments on the unit's performance throughout all stages of the exercise. Evaluators monitor the critical point of the action but must avoid interfering. They must not disrupt unit performance or undermine realism. They must be trained and given standardized evaluation guides if their feedback is to be meaningful; MTPs and T&EOs are a good start. Evaluators facilitate AARs alone or along with the unit leader.

TACTICAL ENGAGEMENT SYSTEM

A-18. The TES realistically replicates the effects of actual fire in training. It uses laser adapters on the weapons and laser-sensitive equipment on the Soldiers and vehicles. A TES weapon emits a laser beam when fired. Accurate fire strikes the laser-sensitive equipment and scores a hit. The TES requires Soldiers to aim accurately and to use cover and concealment effectively. When used with an OPFOR, TES offers realistic training. It is an excellent training aid for an FTX or STX. It is also effective at the shoot house as a substitute for live fire. The two major shortcomings of TES are that it allows Soldiers to seek cover behind structures that normally would not protect them, and that it does not replicate suppression.

NAVIGATION

A-19. Navigational techniques for urban areas differ from those needed on other terrain.

- A typical 1:50,000-scale military map provides insufficient detail for accurate navigation. Therefore, Soldiers must learn to navigate using commercial maps, sketches, and aerial photographs. Leaders must learn how to plan operations using these items.
- Underground navigation is a special challenge. Soldiers must know subway plans and be able to identify tunnels, elevated tracks, stairways, and other critical features. They must learn how to measure distance, remain oriented, and mark their routes.
- Destruction often erases urban features that could aid in navigation. Soldiers must learn to orient using distinctive, difficult-to-erase features such as bridges, waterways, towers, and intersections.
- Leaders conduct navigational exercises in an actual urban area. Soldiers must move from point to point while controllers at each point monitor and verify Soldiers' progress.
- The GPS is a navigational aid that triangulates with satellites to calculate position. However, this system may be affected by nearby tall buildings, and its effectiveness decreases inside buildings or underground.

LIMITED VISIBILITY TRAINING

A-20. Urban operations involve day and night operations. FTXs, STXs, CFXs, and CPXs are the most valuable exercises for night training.

- Field-expedient lighting, NVDs, illumination, and electrical lighting must be exploited when the tactical situation allows. Soldiers must learn discipline in their use of light, and also learn the lighting techniques that allow them to carry out their urban training program.

- Situational awareness is the Soldier's knowledge of his location, the locations of friendly forces, and the locations of the enemy on the battlefield. Situational awareness is the key to avoiding fratricide, especially during limited visibility operations. Units must plan to use all the equipment and devices available to mark friendly units and to avoid fratricide, without exposing themselves to enemy observation.
- Reverse-cycle training should be part of an urban training program. Limited visibility fighting skills are perishable, but they can be sustained and improved by effective training. The goal is for every Soldier to operate as effectively at night as during the day.

SAFETY AND AMMUNITION CONSIDERATIONS

A-21. When using ammunition and pyrotechnics, all associated safety requirements must be followed. This paragraph describes some of the safety equipment and training procedures to consider:

SPECIAL EFFECT SMALL-ARMS MARKING SYSTEM

A-22. SESAMS is 5.56-mm and 9-mm color-marking training ammunition. It is designed for close combat training. It is used with the close combat mission capability kit (CCMCK) adapter kit, which is installed on the M4/16 rifle, M9/M11 pistol, and M249 SAW. Different colors can be assigned to OPFORs for engagement identification and fratricide tracking. SESAMS has a 30-meter effective range and activates TES (MILES), so the two are used together. Anytime during a SESAMS exercise that a Soldier loses any of his protective gear, he must immediately cover his face and eyes with his hands. He must stay put until the engagement ends, or until he is told that it is safe to replace his safety equipment. Everyone, to include observer-controllers, must wear the protection shown in Table A-2.

Table A-2. SESAMS protection.

SKIN	Wear gloves, and cover all exposed skin. At close ranges, the SESAMS projectile can break skin.
HEAD	Take every precaution to protect your head.
EYES	Keep your eyes completely enclosed within eye protection. You can use standard Army sun, wind, and dust goggles.
EARS	Wear hearing protection to prevent hearing loss.
MOUTH	Keep your mouth covered.

DANGER

Failure to wear the required protection during SESAMS exercises can result in injury or death.

M84 STUN GRENADE

A-23. The hand-deployed M84 stun grenade serves as a reliable, effective, nonlethal means of neutralizing, diverting, and disorienting enemy personnel. It is used tactically as well as in training.

A-24. The M84 produces an intense (1 to 2.5-million candlepower) flash and, at 1.5 meters, a 170 dBs bang. Even though this is too low to produce permanent hearing damage, one-time use Soldiers are required to wear hearing protection anytime noise levels exceed 85 dBs.

A-25. The M84 stun grenade has an M201A1 fuze and a pyrotechnic output charge in a synthetic casing with a one- to two-second delay fuze, which cannot be cooked off. The grenade body has a nonfragmenting metal hexagonal casing, which, once deployed, resists rolling.

SMOKE GRENADES

A-26. Smoke grenades are used in urban combat to cover movement, deceive enemy forces as to the exact location of the attacking force, and for signaling. Smoke grenades should not be used in buildings and subterranean passages as they displace oxygen and protective masks do not filter the smoke or provide oxygen. Buildings in urban areas channel air currents, which influence emplacement of smoke grenades. One grenade should be used to test the airflow before employing additional grenades.

WARNING

Avoid using smoke grenades in buildings and in subterranean passages.

OPPOSING FORCES

A-27. OPFOR should be used as often as possible to enhance the realism of training, especially for FTXs, CFXs, and STXs. Thebe realistic, the OPFOR must have defined countertasks. These countertasks must support "Blue Force" training, and should prevent accidental contact between the Blue Force and the OPFOR. Leaders should develop a thorough control plan for OPFOR scenarios. This plan should describe uniform and vehicle markings, ammunition, safety instructions, controller guidance, and guidelines for handling enemy prisoners of war (EPW) and noncombatants. The ROE must clearly specify the limits of close combat between Soldiers.

- OPFOR heighten Soldiers' interest by giving them a real opponent. This fosters unit competition and realism, especially when the OPFOR is used in addition to TES and SESAM.
- OPFOR challenges leaders' ability to improvise. However, this free-play scenario should be monitored to ensure the unit meets the training objectives.
- OPFOR can use threat urban operations and doctrine to determine their course of action. If available, OPFOR uniforms, mock documents, overlays, and equipment add realism and intelligence to the scenario. Leaders conduct preliminary training for the OPFOR, which helps the OPFOR recreate the threat's urban tactics.
- Noncombatants should be integrated into training scenarios.

PRECISION MARKSMANSHIP TRAINING

A-28. Precision marksmanship training should be conducted before training at the UAC, CACTF, and the Shoot House, which accommodate training in target discrimination (FM 3-22.9). This is the best use of the unit's time and resources.

PHYSICAL READINESS TRAINING

A-29. Combat in urban areas places extraordinary physical demands on Soldiers. Therefore, physical readiness training, with emphasis on upper body development, is paramount to any urban operations training program.

VARIED SCENARIOS

A-30. Leaders must vary training scenarios to prevent redundant and predictable tactical behavior. Using predictable avenues of approach into the facility produces stereotypical tactical solutions.

BUILDING FACADE WALLS, WINDOWS, AND DOORS

A-31. The use of façade walls, windows and doors, in conjunction with SESAMS in the interior of some stations at the UAC, provides for more realistic training. This method of training reinforces the fact that walls, windows, and doors in most cases *do not* stop the penetration of service ammunition. Plastic wrap sheeting, tar paper, or some other type of material can be used to allow SESAMS penetration (Figure A-1).

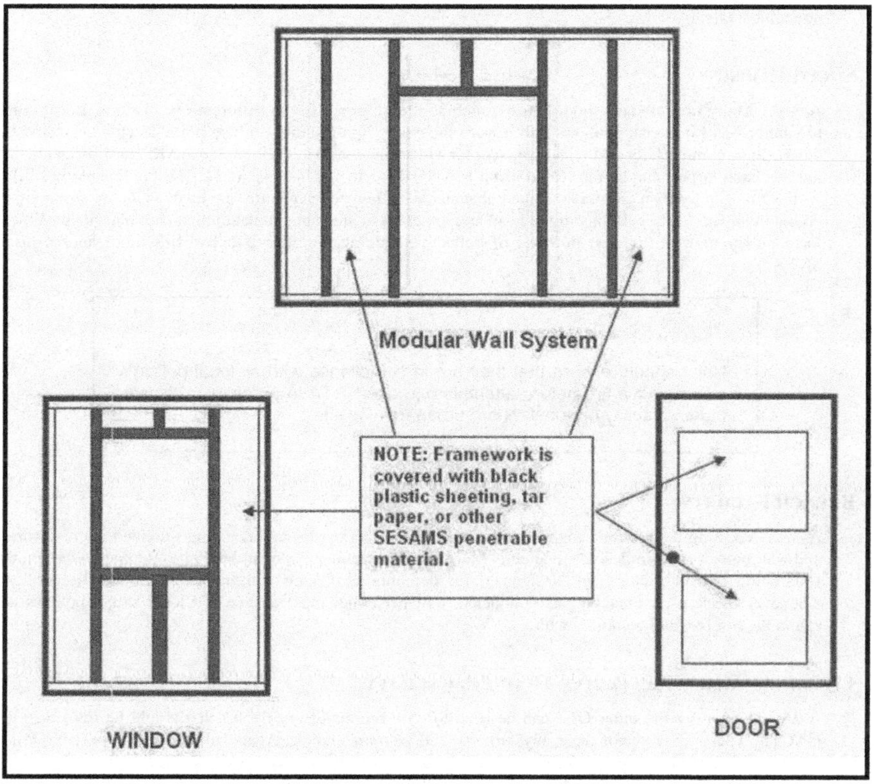

Figure A-1. Shoot-through walls and doors.

USE OF OLDER OR NONSTANDARD FACILITIES

A-32. Until installations receive funding for the UAC, SH, BF, and CACTF, they can use the older MOUT Assault Course (MAC) and Collective Training Facility (CTF) as described in the 2002 edition of TC 90-1, which this edition supersedes. Installations may also use nonstandard shoot houses and other nonstandard urban facilities. These facilities may be adapted as follows:

URBAN ASSAULT COURSE

A-33. Generally, the MAC can be used to train most of the UAC tasks. Of course, the MAC lacks targetry and instrumentation. Instead, units can record training with portable video cameras. During the AARs, they can play back the videos on VCRs hooked to TVs or on TV/VCR combos. OPFOR combatants will have SESAMS or MILES. People in civilian clothes can play the parts of the noncombatants.

SHOOT HOUSE

A-34. Many nonstandard shoot houses, such as tire houses, exist in different installations. Local range regulations and unit risk analyses will govern their use. Again, units can use portable video cameras for video capture and VCRs and television sets for video play back to facilitate an AAR. Both precision and nonprecision targets can be fabricated using E-type silhouettes on a stake, old BDUs/civilian clothes filled with straw, and so on. Simulate noncombatants with E-type silhouettes painted white or dummies in civilian clothes. Place balloons in the head and upper torso areas to simulate lethal shot placement. Where local policy permits, convert portions of collective training facilities into live-fire shoot houses using SRTA.

CAUTION

Units should ensure that they are in compliance with all local policies affecting live fire before attempting to use SRTA in an older collective training facility or nonstandard urban training site.

BREACH FACILITY

A-35. Existing demolition ranges may be used to conduct explosive breaching provided door, window, and wall units, or facsimiles, are available. Mechanical breaching can be trained in cantonment areas on the installation where building are designated for demolition, if such facilities are available. Regardless, whenever mechanical breacking is conducted, unit personnel must ensure that local safety policies are available, briefed, and complied with.

COMBINED ARMS COLLECTIVE TRAINING FACILITY

A-36. Generally the older CTF can be used to train the majority of tasks that would be trained in the CACTF. The CTF will not have any targetry and instrumentation. Again, units can use portable video cameras for video capture and VCRs and television sets for video play back to facilitate an AAR. Targets can be OPFOR with MILES. Role players in civilian clothes can be introduced to play the part of noncombatants. If armored vehicles are used, units should ensure that local range policies allow for their use. Older CTFs are normally either a 16 or 32 building design. Appropriate unit-level training, platoon, company, or battalion, should be planned based on the size of the facility.

Appendix B
Unit Training Matrix

Unit training plans are prepared IAW FM 7-1, which states that "Planning links the unit METL and the execution of battle-focused training.". The commander begins his planning with guidance from a higher echelon and an in-depth assessment of METL tasks (in this case, mission-essential urban tasks). Based on his own observations, feedback from his subordinates, and other sources the commander determines which tasks should be trained, which should be trained first, and how much time should be allotted for each. Tables B-1 through B-5 list training tasks by the WFF to be trained at each facility.

Table B-1. Army unit training matrix, brigade tasks.

Brigade Tasks	*SHOOT HOUSE Semiannual*	*BREACH HOUSE Semiannual*	*UAC Quarterly*	*CA/CTF Semiannual*	*CTC *As Scheduled*	*SIMULATION Annual*
Develop Intelligence						
Perform Intelligence Operations					X	X
Perform S-2 Operations					X	X
Deploy/Conduct Maneuver						
Attack					X	X
Perform Air Assault					X	X
Defend					X	X
Lodgment					X	X
Exercise Command and Control						
Command and Control the Brigade					X	X
Maintain Communications					X	X
Coordinate Air Defense Operation					X	X
Conduct Noncombatant Evacuation					X	X
Develop a Media Plan					X	
Conduct Mediation and Negotiation					X	
Employ Firepower						
Execute Fires					X	X
Employ Air Support					X	X
Employ Attack Helicopters					X	X
Synchronize Fire Support					X	X
Synchronize CAS					X	X
Conduct Targeting Process					X	X
Protect the Force						
Perform Mobility/Survivability Operations					X	X
Provide Engineer Support					X	X
Coordinate CBRN Operations					X	X
Conduct Operations Security					X	X
Perform Sustainment						
Evacuate Casualties					X	X
Treat Casualties					X	X
Sustain the Force					X	X
Conduct a Civil-Military Operation					X	X
Provide Humanitarian Support					X	X

Table B-2. Army unit training matrix, battalion tasks.

Battalion Tasks	*SHOOT HOUSE Semiannual*	*BREACH HOUSE Semiannual*	*UAC Quarterly*	*CA/CTF Semiannual*	*CTC As Scheduled*	*SIMULATION Annual*
Develop Intelligence						
Conduct S-2 Operations				X	X	X
Deploy/Conduct Maneuver						
Attack a Built-Up Area				X	X	X
Plan for Urban Operation				X	X	X
Defend an Urban Area				X	X	X
Conduct an Infiltration/Exfiltration				X	X	X
Perform Cordon and Search Operations in an Urban Area				X	X	X
Plan for Urban Operations				X	X	X
Conduct Tactical Movement				X	X	X
Conduct Presence Operation in a Stability Environment				X	X	X
Defend an Urban Area				X	X	X
Traverse a Built-up Area				X	X	X
Conduct a Mobile Defense				X	X	X
Employ Firepower						
Integrate Fires and Effects Support				X	X	X
Establish the Fire Support Element				X	X	X
Protect the Force						
Conduct Mobility or Survivability Operations				X	X	X
Secure Civilians During Operation				X	X	X
Conduct Passive Air Defense Measures				X	X	X
Conduct Active Air Defense Measures				X	X	X
Employ Operations Security Measures				X	X	X
Integrate Air Defense Support				X	X	X
Perform Sustainment						
Conduct S-1 Operations				X	X	X
Conduct S-4 Operations				X	X	X
Conduct a Civil-Military Operation				X	X	X
Exercise Command and Control						
Conduct S-6 Operations				X	X	X
Conduct Noncombatant Evacuation				X	X	X
Conduct S-3 Operations				X	X	X
Conduct Mediation and Negotiations				X	X	X
Plan Roadblocks or Checkpoints				X	X	X
Establish Liaison				X	X	X

Table B-3. Army unit training matrix, company tasks.

Company Tasks	*SHOOT HOUSE* *Semiannual*	*BREACH HOUSE* *Semiannual*	*UAC* *Quarterly*	*CA/CTF* *Semiannual*
Develop Intelligence				
Conduct an Area or Zone Reconnaissance				X
Reconnoiter a Built-Up Area				X
Deploy/Conduct Maneuver				
Conduct Cordon and Search in a Built-Up Area				X
Attack a Built-Up Area				X
Conduct Presence Operations				X
Conduct Tactical Movement in a Built-Up Area				X
Defend in Urban Environment				X
Conduct an Infiltration (Co/Plt)				X
Conduct Operations with Armored or Mechanized Infantry Vehicles in an Urban Environment				X
Conduct Civil Disturbance Operations				X
Establish Checkpoints				X
Employ Firepower				
Integrate Direct Fires				X
Integrate Indirect Fires				X
Protect the Force				
Maintain Operations Security				X
Conduct Passive Air Defense Measures				X
Conduct Passive Air Defense Measures				X
Perform Sustainment				
Treat Casualties				X
Evacuate Casualties				X
Secure Civilians during Operations				X
Conduct Resupply Operations				
Process Captured Documents and				X
Equipment				X
Handle Enemy Prisoners of War				
Exercise Command and Control				
Prepare for Combat				X
Establish Communications				X
Conduct Consolidation and Reorganization				X
Conduct Negotiation				X
Conduct Troop-Leading Procedures				X

Table B-4. Army unit training matrix, platoon/squad tasks.

Platoon/Squad Tasks	*SHOOT HOUSE* Semiannual	*BREACH HOUSE* Semiannual	*UAC* Quarterly	*CA/CTF* Semiannual
Develop Intelligence				
Report Tactical Information	X		X	X
Conduct an Area or Zone Reconnaissance	X		X	X
Reconnoiter a Built-Up Area	X		X	X
Deploy/Conduct Maneuver				
Conduct Tactical Movement in a Built-Up Area	X		X	X
Assault a Building	X		X	X
Conduct a Strongpoint Defense of a Building				X
Conduct an Infiltration or Exfiltration				X
Conduct Operations with Armored or Mechanized Vehicles in an Urban Environment	X			X
Conduct a Presence Patrol	X			X
React to a Civil Disturbance				X
Establish a Checkpoint				X
Search a Building	X		X	X
Employ Firepower				
Employ Fire Support				X
Protect the Force				
Maintain Operations Security				X
Conduct Passive Air Defense Measures				X
Conduct Active Air Defense Measures				X
Perform Sustainment				
Secure Civilians During Operations	X		X	X
Perform Resupply Operations				X
Treat Casualties	X		X	X
Evacuate Casualties	X		X	X
Handle Enemy Prisoners of War				
Exercise Command and Control				
Conduct Troop-Leading Procedures	X		X	X
Establish Radio Communication	X		X	X
Conduct Consolidation and Reorganization	X		X	X
Prepare for Combat	X		X	X

Table B-5. Army unit training matrix, individual tasks/techniques.

Individual Tasks/Techniques	*SHOOT HOUSE* *Semiannual*	*BREACH HOUSE* *Semiannual*	*UAC* *Quarterly*	*CA/CTF* *Semiannual*
Skill Level 1				
Perform Movement Techniques in an Urban Environment, 071-326-0541	X	X	X	X
Prepare Positions for Individual and Crew-Served Weapons in an Urban Environment, 071-326-0550	X	X	X	X
Select Hasty Fighting Positions in an Urban Environment, 071-326-0557	X	X	X	X
Employ Hand Grenades, 071-325-4407	X	X	X	X
Engage Targets with an M16-Series or M16A2 Rifle, 071-311-2007	X	X	X	X
Execute a Mechanical Breach, 071-440-0027			X	X
Skill Level 2				
Conduct an Explosive Breach	X	X	X	X
Conduct a Breach, 071-440-0029	X	X	X	X
Prime Explosive Nonelectrically, 051-193-1003	X	X	X	X
Construct a Nonelectric Initiating Assembly with Modernized Demolition	X	X	X	X

Appendix C

Stability and Support Operations Unit Training Matrix

When assigned a stability or support mission, a well-trained unit must be able to rapidly shift its focus from war fighting to stability and support and also from stability and support to war fighting. During a stability or support operation, the unit performs numerous activities. Essentially, the unit accomplishes these activities through execution of tactical missions and tasks. While stability and support operations can occur anywhere, it is most probable that they will occur in an urban environment. Tables C-1 through C-4 list training tasks by the WFF to be trained at each facility.

Table C-1. Stability and support operations unit training matrix, brigade tasks.

Brigade Tasks	*SHOOT HOUSE*	*BREACH HOUSE*	*UAC*	*CA/CTF*	*CTC*	*SIMULATION*
Develop Intelligence						
Collect and Disseminate Information					X	X
Deploy/Conduct Maneuver						
Demonstrate a Show of Force					X	
Employ Psychological Operation					X	X
Employ Firepower						
Provide Fire Support					X	X
Protect the Force						
Protect the Force					X	
Perform Sustainment						
Restore Law and Order					X	
Exercise Command and Control						
Establish an Evacuation Control Center/Conduct Evacuation Control Center Operations					X	
Negotiate					X	
Perform Noncombatant Evacuation Operations					X	
Plan a Media Visit					X	
Develop Intelligence						
Collect and Disseminate Information				X	X	X

Table C-2. Stability and support operations unit training matrix, battalion tasks.

Battalion Tasks	*SHOOT HOUSE*	*BREACH HOUSE*	*UAC*	*CA/CTF*	*CTC*	*SIMULATION*
Deploy/Conduct Maneuver						
Interdict Smuggling Operations				X	X	
Disarm Belligerents				X	X	
Handle Captured Insurgents or Belligerents				X	X	
Employ Psychological Operation				X	X	
React to Civil Disturbance				X	X	
Employ Firepower						
Provide Fire Support				X	X	X
Protect the Force						
Protect the Force				X	X	
Restore Law and Order						
Identify and Process Detainees				X	X	
Deliver Supplies or Humanitarian Aid				X	X	
Control Civilian Movement				X	X	
Prepare Traffic Control Plan				X	X	
Exercise Command and Control						
Negotiate				X	X	
Perform Noncombatant Evacuation Operations				X	X	
Conduct Marshalling Force Operations				X	X	
Negotiate a Belligerent Force Checkpoint				X	X	
Plan a Media Visit				X	X	
Perform a Cordon and Search				X	X	
Coordinate and Monitor Convoy Security				X	X	

Table C-3. Stability and support operations unit training matrix, company tasks.

Company Tasks	SHOOT HOUSE	BREACH HOUSE	UAC	CA/CTF	CTC	SIMULATION
Develop Intelligence						
Collect and Disseminate Information				X	X	
Deploy/Conduct Maneuver						
Interdict Smuggling Operations				X	X	
Disarm Belligerents				X	X	
Handle Captured Insurgents or Belligerents				X	X	
Defend Convoy				X	X	
Apprehend/Detain Noncombatants				X	X	
Perform Cordon and Search				X	X	
Employ Firepower						
Provide Fire Support				X	X	
Protect the Force						
Establish a Checkpoint				X	X	
Operate a Checkpoint				X	X	
Protect the Force				X	X	
Perform Sustainment						
Process Captured Documents and Equipment				X	X	
Exercise Command and Control						
Negotiate a Belligerent Force Checkpoint				X	X	
Conduct Marshaling Force Operations				X	X	
Negotiate a Belligerent Force Checkpoint				X	X	
Plan a Media Visit				X	X	
Coordinate and Monitor Convoy Security				X	X	
Link Up with a Convoy				X	X	

Table C-4. Stability and support operations unit training matrix, platoon/squad tasks.

Platoon/Squad Tasks	*SHOOT HOUSE*	*BREACH HOUSE*	*UAC*	*CA/CTF*	*CTC*	*SIMULATION*
Deploy/Conduct Maneuver						
Disarm Belligerents				X	X	
Handle Captured Insurgents or Belligerents				X	X	
Apprehend/Detain Noncombatants				X	X	
Search a Building	X			X	X	
React to a Sniper			X	X	X	
Conduct Platoon Riot Control Formation				X	X	
Protect the Force						
Establish a Checkpoint				X	X	
Operate a Checkpoint				X	X	
Protect the Force				X	X	
Exercise Command and Control						
Negotiate a Belligerent Force Checkpoint				X	X	
Exercise Command and Control						
React to a Media Interview				X	X	

Appendix D

Leader Development Training

To be successful in combat, the Army must continually train to develop and maintain combat-ready Soldiers, leaders, and units that can perform assigned tasks to specific standards.

SIMULATIONS TRAINING

D-1. Commanders train their staffs to plan, coordinate, and synchronize their units' operations from initial mobilization and deployment. These operations include the full spectrum of Army operations; offense, defense, stability, and support. Current simulations lack adequate replication of urban scenarios, terrain, and weapons effects. The introduction of WARSIM 2000 and OneSAF should help to solve these simulations modeling deficiencies. For lower echelon units, battalion and below, constructive simulations complement collective battle staff training that occurs during virtual simulation similar to the close combat tactical trainer and live events. For echelons above battalion, it becomes difficult to use simulators or live maneuver to train. Studies to date indicate the crossover point for simulation versus simulator is at the battalion level. Some training events from battalion and below can be trained effectively using various types of simulators. Above battalion level, effective training almost has to be conducted with a simulation (Figure D-1).

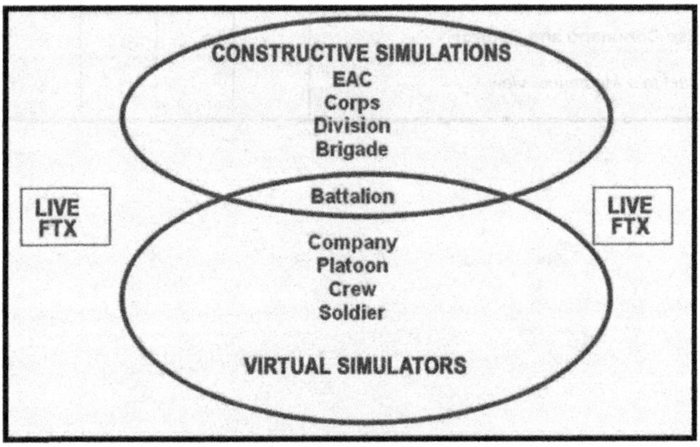

Figure D-1. Echelons and uses.

D-2. Simulations that support leader and battle staff training focus on collective tasks rather than individual leader tasks. Simulations, used in this context, rely on the human dynamics in the command setting and use real-time actions in a tactical environment with a unit's go-to-war systems. This process forces commander and staff interaction that provides feedback and helps refine the integration process. Battle staff training encourages team building, task coordination, and standardization of staff actions.

- Constructive simulations can provide commanders large areas for conducting training, access to sensitive areas through computer and mapping technology, and futuristically provide a much better level of fidelity in terms of urban operations terrain and scenario replication.
- Simulations can portray large, capable, and doctrinally correct OPFORs. All of the Soldiers portrayed in the units in a simulation do not have to be present in order for the training to be realistic. An attractive feature of simulations has always been the reduced staffing necessary to conduct this type of training exercise.
- Simulations can stress CPs and provide the realistic conditions under which commanders must make decisions. Brigade and below simulation (BBS) and corp battle simulation (CBS) are well known for the stressful environments created during an exercise. The stress is realistic and can assist in conditioning the training audience before CTC rotations.
- Simulations allow different units to train under the same conditions and to the same standards. This technique is a unique feature of simulations as the exact weather, opposing force organization, location of units, and equipment can be used to train many different units on the same terrain with the same conditions.
- The size and number of personnel in a battle staff vary depending on the echelon and other demands unique to that commander. However, the WFF are represented by the following staff officers: intelligence, operations, fire support, communications, logistics, air defense, and engineer.
- Characteristics and benefits of simulations used to support battle staff training follow:
  - -- Build and sustain staff teamwork.
  - -- Build and sustain effective and efficient organizations.
  - -- Provide training and validation of command and staff procedures (SOPs, TACSOPs, FSOPs).
  - -- Provide proficiency to the individual staff officer, while building proficiency in a fully integrated staff.
  - -- Allow practice of battle drills and tactical missions (movement to contact, hasty defense).
  - -- Allow commanders to determine if their staff and subordinate commanders understand the commander's intent and concept of operations.

D-3. The major development efforts for FAMSIM is WARSIM 2000. WARSIM 2000, in conjunction with OneSAF, exploits new technologies to enable commanders and staffs at battalion through EAC to train in a realistic simulation environment (Table D-1). The increased realism of WARSIM 2000 over existing models allows units to use Army Battle Command Systems (ABCS) to synchronize across the complete range of operations. WARSIM 2000's design allows warfighting CPs to use organic ABCS equipment to interact with the simulation from training sites in the field, as well as from simulation centers. WARSIM 2000 will also be capable of showing a joint and combined environment across the operational continuum.

Table D-1. Urban operations simulations crosswalk.

CAMTF Training Tasks	*WARSIM*	*OneSAF*
Corps	X	
Division	X	
Brigade		X
Battalion		X
Company		X
Platoon		X

D-4. Future simulation capabilities simultaneously integrate live, virtual, and constructive environments. Constructive simulation is a viable solution for battalion and above training and the federation of WARSIM and OneSAF models should alleviate the following training simulation problems:

- Eliminate the need for swivel chair interface.
- Allows focus on subordinate units fighting the fight.
- Digital and analog feeds to C4I systems including FBCB2.
- May provide for distance learning opportunities.
- Facilitates simultaneous multi-echelon urban operations training

D-5. At company and below, the best training solution is live and virtual. This is accomplished by linking the Close Combat Tactical Trainer (CCTT) to OneSAF and VLET to Engagement Skills Trainer (EST) technology. Entity-based models like OneSAF can allow the leader to fight from a TOC or vehicle, if linked to virtual or live environments. It also has the potential for distance learning.

PHASES OF TRAINING

D-6. The Army Urban Operations Training Strategy consists of initial, intermediate, and culminating stages. These phases of training, shown in Figure D-2, describe the sequence in which urban operations training tasks should be conducted.

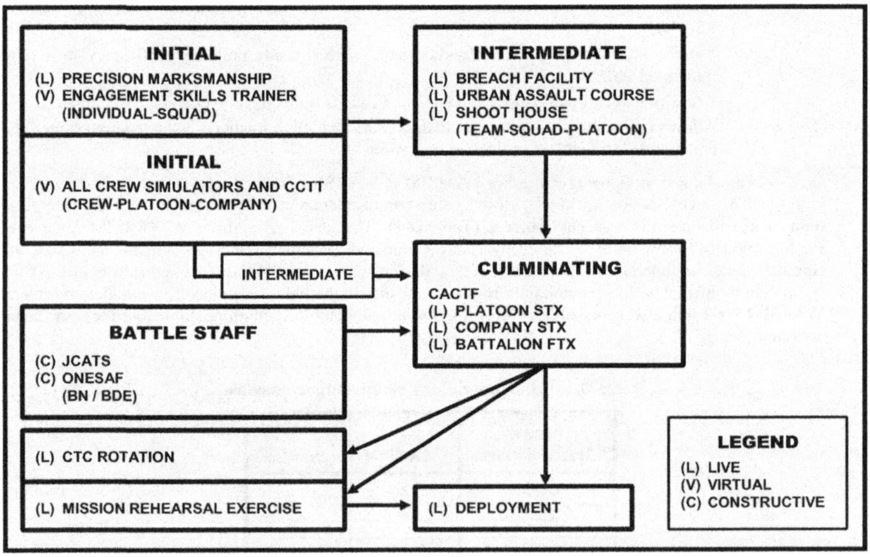

Figure D-2. Phases in the Live, Virtual, and Constructive Training Strategy.

Initial Phase

D-7. The initial phase of training covers those individual and collective tasks that Soldiers and leaders should be proficient at before arriving at any of the urban operations training sites. They consist primarily of basic marksmanship skills and Soldier tasks needed to perform any tactical operation.

Intermediate Phase

D-8. The intermediate phase consists of training conducted at the breach facility, shoot house, and the urban assault course. Urban operations training tasks should be conducted here before advancing to the Combined Arms Collective Training Facility with the exception of Aviation, Armor, and other branches that have no practical use for the Shoot House, Breach Facility, or Urban Assault Course.

Culminating Phase

D-9. The combined arms collective training facility is the culminating phase of urban operations training and focuses on multiechelon collective skills for all branches of the combined arms team. Urban operations leader training currently exists in our leader professional development courses. Units have developed internal urban warfare training programs and TRADOC schools have assembled technical mobile training teams (MTTs) for marksmanship, combatives, and breaching techniques.

Education Phase

- Urban warfare reading/video list.
- Urban operations equipment.
- Review of doctrinal publications (FM 3-21.10, FM 3-06.11, for Infantry) and corresponding publications for other branches.
- Small-unit urban movement techniques and considerations.

Planning and Execution Phase

- Prepare company OPORD or WFF-related products.
- Brief platoon/company OPORD/WFF-related products to commander or staff representative.
- Participate as briefer in urban environment TEWT

Sustainment Phase

- Participate in company/platoon TEWT or other urban operations event
- Execute/facilitate a company or platoon TEWT.

EXAMPLE LEADER'S COURSE POI

D-10. An example urban leaders' course POI, which is intended to provide unit trainers with a tailorable urban operations training program, is shown in Table D-2. It is a *train-the-trainer* course for team leaders through platoon leaders to train tactics, techniques, and procedures of advanced urban operations skills. The length and focus of the POI can be tailored to support the commander's training objectives and unit METL. The target audience is newly arrived, or soon to be, small unit leaders. The recommended course length is about two weeks or 120 to 140 hours of instruction.

REALISTIC TRAINING OFF FEDERAL FACILITIES

D-11. Realistic training for urban operations is critical to force readiness. Training opportunities that are both realistic and challenging may be unavailable at federal installations, or at facilities specifically designed for urban operations training. Training in off-post civilian urban settings may become necessary in order to support readiness and training requirements.

D-12. The Department of Defense has developed the following SOP for planning and approving such events. The SOP assists in balancing the needs of the military training and of the civilian community/ies to be used for realistic urban training events.

Table D-2. POI for urban leader's train-the-trainer course.

Course Overview	• Introduction	
Precision Marksmanship	• Firing techniques. • Target discrimination.	
Breaching Techniques	• Explosive. • Ballistic. • Mechanical.	

Table D-2. POI for urban leader's train-the-trainer course (continued).

Offensive Operations	• Planning considerations	-- METT-TC. -- Snipers. -- CASEVAC. -- Communications. -- ROE.
	• Platoon attack a built-up area (task organization).......................................	-- Assault (isolate, enter, clear). -- Rooms, hallway, stairs, superstructure, subsurface. -- Markings. -- Movement techniques. -- Streets. -- Alley ways. -- Obstacles. -- Other danger areas. -- Open areas. -- Roof tops.
	• Combined arms	-- Armor/mechanized platforms. -- Aviation assets. -- Artillery. -- Engineer.
	• Limited visibility	
Defensive Operations	• Planning considerations	-- METT-TC. -- Snipers. -- CASEVAC. -- Communications. -- ROE.
	• Characteristics of urban areas	
	• Hasty defense	-- Occupation and preparation of positions. -- Hasty firing positions. -- Improving the defense.
	• Deliberate defense, including priorities of work and other considerations	
	• Defensive plan at platoon level.	-- Defense of a strongpoint. -- Defense against armor.
	• Combined arms	-- Armor/mechanized. -- Aviation assets. -- Artillery. -- Engineer.
	• Limited visibility • Fundamental combat skills	

Table D-2. POI for urban leader's train-the-trainer course (continued).

Stability and Support Operations	• Principles of stability and support.	
	• Stability and support activities	-- NEO. -- Humanitarian assistance. -- Disaster relief. -- Show of force. -- Attack and raids.
	• Rules of engagement application • Convoy operations • Checkpoints • Fire support • Sniper employment/counter sniper measures • Civil disturbance	
Urban Training Facilities	• Breach facility • Urban assault course • Shoot house • CACTF	
New Technologies	• Rifle launch entry munition	
	• Unmanned vehicles	-- UAV. -- UGV. -- Ladders. -- Night vision devices.

STANDING OPERATING PROCEDURES FOR PLANNING AND APPROVING REALISTIC URBAN TRAINING

1. **REFERENCES.**

   10 USC § 167, *Unified Combatant Command for Special Operations Forces.*

   10 USC §§ 3013, 5013, 8013.

   Department of Defense Directive 5111.10.

   Unified Command Plan 99.

2. **PURPOSE**. This SOP establishes uniform planning and approval procedures for realistic urban training (RUT) events.

3. **APPLICABILITY**. This SOP applies to all RUT events in civilian urban settings within the United States, its territories, and its possessions; and to Active or Reserve Forces, including National Guard Forces in Federal service. This policy does not apply to National Guard Forces under in-state status (Title 32), or to aviation operations that do not require an exemption from standard Federal Aviation Regulations. The training event should not be located on federal state, local, or private installations or facilities.

4. **DEFINITIONS**. For the purposes of this SOP, the following definitions apply:

a. **Urban Operations. This term** includes (see FM 3-06) all military actions planned and conducted on a topographical complex, and on adjacent natural terrain, where man-made construction is the dominant feature. It includes combat in towns and cities, including house-to-house and street-by-street fighting.

b. **Realistic Urban Training.** RUT is high-intensity, close-quarter battle training using live or nonlethal fires; demolitions or explosives such as breaching charges; and air and naval supporting platforms at the objective, in ***civilian urban settings***.

c. **Civilian Urban Setting.** A civilian urban setting is any area(s) located in or near a civilian community. Civilian urban settings will not be developed or established for training events, especially where training activities would affect civilian residences or commercial areas due to noise, traffic, and so on.

5. **PROCEDURES**. The following procedures apply to the planning and approval of RUT in civilian urban settings:

a. The combatant commander or the service ensures compliance with this SOP. Approval authority for RUT events may be delegated to a general or flag officer within the chain of command, or to the combatant common service conducting the training.

b. The approval authority reviews proposed training events to determine whether they are necessary and appropriate for unit readiness. Training must support unit METs and JMETs.

c. On approval of the proposed training, and at least 14 days before training starts, the combatant common service notifies the **SECR/PF** through the CJCS (by way of message) of the approved training. At a minimum, this notice includes the following:

- Name and location of the unit conducting training.
- Scope of training and activities planned such as live fire, breaches, or use of aircraft.
- Start and end dates.
- Training location.
- A statement that all necessary coordination has been accomplished IAW this SOP. The statement should include a list of the contacted official's names and their positions.

d. **Coordination with Civilian Officials.**

- While planning the training event, the commander coordinates with and obtains written approval to conduct the event from the appropriate federal, state, and local civilian authorities. Commanders will consult their legal and public affairs officers to identify the civilian officials, for example the mayor, borough chief, county commissioners or supervisors, with whom they must coordinate the training event. To do this, they consult local and Federal law enforcement agencies. At a minimum, they must consult a senior official with responsibility for each affected civilian urban setting.
- During coordination, the commander covers the details of the proposed training activities, such as whether the event involves live fire, rotary- or fixed-wing aircraft, or close quarters battle training; and whether the exercises will be conducted during the day, night, or both: and so on. It will also list actions to ensure public and participant safety, proposed public or resident notification and outreach procedures, and any other issues that might cause public interest or concern.
- As much as possible, the commander briefs civilian officials in person on the exercise concept.

- The US congressional representative or staff for the training district, and the US senators for the state, or their staffs, are advised of the proposed training event and given whatever information they request. Neither written approval nor concurrence from the representative or senators is required. However, if these officials raise objections or issues, and if those issues cannot be resolved, they are immediately elevated to the Joint Staff, J3, and the service or combatant commander point of contact.
- For each training site, license agreements with civilian or government property owners or managers are executed. Agreements address liability and conditions of acceptable usage of property. Property owners are informed of the nature of the training including written notification as to whether the activities will include use of breaching charges and, if so, how much; or live ammunition and, if so, which weapons and what ammunition. This notice will also state *if* rotary or fixed-wing aircraft will be used *and* whether the exercises will be conducted during the day, night, or both. Before the exercise can begin, all license agreements must be reviewed for legality. Copies of the license agreements will be provided to the senior civilian officials responsible for training area.
- A record of coordination activities is maintained for each training event. The record includes the names and positions of the officials *who* coordinate and approve the event. It also provides officials with a summary of the events such as date, time, and location of the training; types of activities; and planned public notification and outreach. A follow-on letter is distributed to the senior civilian officials responsible for the affected civilian urban settings. The letter confirms the training event *and* lists the names of the civilian officials who plan the training event.
- The unit may schedule recurring training events at a particular area or location by submitting a single MOA to the authorizing civilian officials. The terms of the MOA must comply with the requirements of this SOP. The service or the combatant commander reviews the MOA, which authorizes the recurring training, to ensure it complies with the procedures. Once the commander approves the MOA, the unit may conduct individual training events under it without further Joint Staff or OSD review. However, services and combatant commanders notify SO/LIC *and* the Joint Staff 14 days *before* the training in this MOA begins. Thereafter, services and combatant commanders must review these MOAs periodically to ensure *that* they continue to fulfil the intent.
- *If* nondisclosure agreements are required, command legal representatives must review them. Nondisclosure statements should clearly explain *that* the agreement does not preclude persons from informing others of general, nonsensitive information about the training exercise.
- Depending on OPSEC and safety considerations, commanders must adhere to the recommendations of civilian officials regarding community notification, outreach, and press procedures. Community notification and outreach procedures may include door-to-door notification by the local law enforcement agency. Commanders will advise civilian officials when they cannot follow their recommendations.

e. Other instructions include--

- Legal and PA representatives attend final planning meetings, and they remain on-site throughout all training activities.
- The on-site PAO handles all press queries after coordinating with the on-scene commander.
- All requested training must follow applicable statutes, law, DOD directives, other guidance, and implementing service directives.

Glossary

Section I. ACRONYMS AND ABBREVIATIONS

1SG	first sergeant
A	
AA	assembly area
AAR	after-action review/report
ABCS	Army Battle Command System
ACU	Army combat uniform
ADA	air defense artillery
ANCOC	advanced noncommissioned officer course
AO	area of operation
AR	Army Reserve
AT4	a shoulder-fired munition, formerly called a "light antiarmor weapon"
atk	attack
ATWESS	antitank weapon effect signature simulator
AVCATT	armored vehicle combined arms tactical trainer
B	
BBS	brigade and below simulation
BCT	brigade combat team
bde	brigade
BDU	(obsolete; *see* ACU)
BF	Breach Facility
bn	battalion
BNCOC	basic noncommissioned officer course
BOLC	Basic Officer Leadership Course
BOS	(obsolete; *see* WFF)
BP	battle position
BSA	brigade support area
BSX	battle simulation exercise
C	
C2	command and control
CACTF	Combined Arms Collective Training Facility
CASEVAC	casualty evacuation
CBRN	chemical, biological, radiological, nuclear
CBS	corps battle simulation
CCMCK	close combat mission capability kit
CCOC	Captain's Career Officer Course
CCTT	close combat tactical trainer
CCTV	closed-circuit television
CFX	command field exercise
CMTC	Citizens Military Training Corps
co	company
commo	communications
CP	command post
CPX	command post exercise
CTC	Combat Training Center
CTF	collective training facility
D	
DES	demolition effects simulator
det cord	detonation cord
div	division
DLIC	detachment left in contact
DS	Digital Training Management System
DTMS	direct support
E	
EA	engagement area
EAC	echelons above corps
ENCATT	engineer combined arms tactical trainer
engr	engineer
EPW	enemy prisoners of war
EST	Engagement Skills Trainer
F	
FAMSIM	family of simulations
FO	forward observer
FRAGO	fragmentary order
FSCATT	fire support combined arms tactical trainer
FSOP	field standing operating procedures
FTX	field training exercise

G

GREM grenade rifle entry munition
GSR ground surveillance radar

H

HC herachloroethane zinc
HEAT high-explosive antitank

I

IAW in accordance with
Inf Infantry

J

JANUS Joint Army-Navy Uniform Simulation
JCATS Joint Conflict and Tactical Simulation
JMET joint mission-essential task
JRTC Joint Readiness Training Center
JTF joint task force
JTX joint training exercise

L

LAW light antiarmor weapon
LFX live-fire exercise
LMG light machine gun

M

MAC MOUT assault course
MACOM major command
MAPEX map exercise
M/C/S mobility/countermobility/survivability
MET mission-essential task
METL mission-essential task list
METT-TC mission, enemy, terrain, troops, time available, and civilians
MI military intelligence
MILES Multiple Integrated Laser-Engagement System
MOPP mission-oriented protective posture
MPAT multipurpose antitank (ammunition)
MPAT-OR multipurpose antitank-obstacle reducing
MRE meals ready to eat
MTT mobile training team

N

NBC (obsolete; *see* CBRN)
NEO noncombatant evacuation operations
NLT not later than
NTC National Training Center

O

OneSAF one semiautomated force (software used to create modeling and simulation programs with military applications)
op operation
OPFOR opposing force
OPORD operation order
OPSEC operations security

P

plt platoon
POI program of instruction
PPP power projection platform
PSP power support platform

R

ROE rules of engagement
ROI rules of interaction
RUT realistic urban training

S

S-1 personnel staff officer (adjutant)
S-2 intelligence staff officer
S-3 operations staff officer
S-4 logistics staff officer
S-5 civil affairs officer
S-6 command, control, communications and computer operations (C4 operations) officer
SAAF small-arms alignment fixture
SBCT Stryker brigade combat team
SDZ surface danger zone
SEP system enhancement program: a version of the M1A2 machine gun with enhanced digital command and control and second-generation forward-looking infrared

SESAMS	Special Effects Small Arms Marking System
SH	Shoot House
SOI	signal operation instructions
SOP	standing operating procedure
sqd	squad
SRTA	short-range training ammunition
STAFFEX	staff exercise
STX	situational training exercise

T

T&EO	training and evaluation outline
TACSOP	tactical standing operating procedure
TES	Tactical Engagement System
TEWT	tactical exercise without troops
TF	task force
TOW	tube-launched, optically tracked, wire-guided (missile)
TTP	tactics, techniques, and procedures

U

UAC	urban assault course
UO	urban operations

V

VCR	videocassette recorder
VLET	virtual leader effects trainer

W

WARSIM	warfighting simulation
WFF	warfighting function
WFX	warfighting exercise

X

XO	executive officer

Section II. TERMS

bustle rack	storage rack mounted on turrets of some armored vehicles
entity-based model	a representation of an actual or conceptual system that includes mathematical, logistical, or computer simulation
warfighting functions	group of tasks and systems (people, organization, information, and processes) united by a common purpose that commanders use to accomplish missions and training objectives. The warfighting functions are intelligence, movement and maneuver, fire support, protection, sustainment, and command and control. These warfighting functions replace the battlefield operating systems (BOSs).

References

SOURCES USED/DOCUMENTS NEEDED

These are the sources quoted or paraphrased in this publication. All of them are needed by and must be available to the intended user.

DOD DIRECTIVE

DOD Directive 2311.01E *DOD Law of War Program.* 9 May 2006.

FIELD MANUALS

FM 3-06	*Urban Operations.* 26 October 2006.
FM 3-06.11	*Combined Arms Operations in Urban Terrain.* 28 February 2002.
FM 3-20.12	*Tank Gunnery (Abrams).* 15 August 2005.
FM 3-21.8	*The Infantry Platoon and Squad.* 28 March 2007.
FM 3-21.9	*SBCT Infantry Rifle Platoon and Squad.* 2 December 2002.
FM 3-21.10	*The Infantry Rifle Company.* 27 July 2006.
FM 3-21.11	*The SBCT Infantry Rifle Company.* 23 January 2003.
FM 3-21.20	*The Infantry Battalion.* 13 December 2006.
FM 3-21.21	*The Stryker Brigade Combat Team Infantry Battalion.* 8 April 2003.
FM 3-21.71	*Mechanized Infantry Platoon and Squad (Bradley).* 20 August 2002.
FM 3-22.9	*Rifle Marksmanship M16A1, M16A2/3, M16A4, and M4 Carbine.* 24 April 2003.
FM 7-0	*Training the Force.* 22 October 2002.
FM 7-1	*Battle Focused Training.* 15 September 2003.
FM 25-4	*How to Conduct Training Exercises.* 10 September 1984.

SOLDIER TRAINING PUBLICATIONS

STP 21-1-SMCT	*Soldier's Manual of Common Tasks Skill Level 1.* 14 December 2007.
STP 21-24-SMCT	*Soldier's Manual of Common Tasks (SMCT) Warrior Leader Skill Levels 2-4.* 2 October 2006.

TRAINING CIRCULAR

TC 7-98-1	*Stability and Support Operations Training Support Package.* 5 June 1997.

Index

120-mm main gun, danger areas, 5-45 (*illus*)

A

Active Army Urban Operations Training Strategy, 1-7 (*illus*)
after-action review (AAR), 1-10
  room, 3-3 (*illus*)
alley wall, securing of, 2-26 (*illus*)
ammunition requirements
  six-day FTX, 5-40 (*illus*)
  Station 4, 2-42 (*illus*)
  Station 5, 2-47 (*illus*)
ammunition safety, A-6
Army National Guard Urban Operations Training Strategy, 1-7 (*illus*)
Army unit training task matrixes
  battalion, B-3 (*illus*)
  brigade, B-2 (*illus*)
  company, B-4 (*illus*)
  individual, B-6 (*illus*)
  platoon/squad, B-5 (*illus*)
Attack scenarios
  FTX, 5-21 (*illus*)
  STX, 5-8 (*illus*)

B

battalion
  integrated training plan, 1-6
  OPORD, 5-23 (*illus*), 5-32 (*illus*)
breach, 2-23 (*illus*)
  door, 4-2 (*illus*)
  wall, 2-25 (*illus*), 4-4 (*illus*)
  window, 4-3 (*illus*)
breach facility, 1-9 (*illus*), 4-1 (*illus*)
  conduct of training, 4-5
  description, 4-2
  safety, 4-7
  training
    and evaluation, 4-7
    audience, 4-3
    information, 4-4
    support requirements, 4-6
building facade, A-8

C

clear
  building, 2-15 (*illus*), 2-43 (*illus*)
  hallway intersection, 2-16 (*illus*)
  next room, 2-18 (*illus*)
  one room, 2-17 (*illus*)
  room by room, 2-17 (*illus*)
  stairs and landing, 2-20 (*illus*)
  wall, 2-28 (*illus*)
collective and supporting individual tasks, related, 3-15 (*illus*)
collective task training, 1-13
Combined Arms Collective Training Facility (CACTF), 1-5 (*illus*), 5-1
  characteristics, 5-1 (*illus*)
  considerations, 5-45
    for training with armor, 5-42
  description, 5-1
  exercises, 5-6
  Infantry rifle
    company FTX Attack, 5-20
    platoon STX Attack, 5-7
  mechanized Infantry
    company or team FTX Defend, 5-29
    platoon STX Defend, 5-12
  purpose, 5-1
  schematic, 5-3 (*illus*)
  tactical exercise without troops (TEWT), 5-6
  target requirements, 5-41 (*illus*)
  targetry, 5-41
command
  and control building, 3-3 (*illus*)
  and staff training, 1-11
company
  FTX, supporting tasks for, 5-20 (*illus*), 5-29 (*illus*)
  OPORD, 3-10 (*illus*)
company/team STX sectors, 5-5 (*illus*)
concurrent training, A-1
conduct of training, 1-10
consolidated support requirements, 3-14 (*illus*)
consolidation and reorganization, 2-32 (*illus*)
Crawl, 1-10, 3-6 (*illus*)

D

danger areas around tank when firing 120-mm main gun, 5-45 (*illus*)
Defend
  Building, 2-43 (*illus*)
  STX, 5-13 (*illus*)
demonstrations, A-2
door, breaching a, 4-2 (*illus*)

E

echelons and uses, D-1 (*illus*)
entry positions, 2-8 (*illus*)
evaluators, A-5
Events
  1, Squad Sequence (*Crawl*), 3-6 (*illus*)
  2, Squad Sequence (*Walk*), 3-7 (*illus*)
  3, Squad Sequence (*Run*), 3-8 (*illus*)
  4, Platoon Evaluation Sequence (*Run*), 3-9 (*illus*)
example scenarios
  Station 1, 2-8 (*illus*), 2-9 (*illus*)
  Station 3, 2-38 (*illus*)
execution of breach, 2-23 (*illus*)
exercises, A-2

F

facade, A-8
facilities, older or nonstandard, A-9
field training exercise (FTX)
  Attack, 5-22 (*illus*)
    battalion OPORD, 5-23 (*illus*)
    FRAGO, 5-27 (*illus*)

field training exercise (FTX)
Attack (*continued*)
Infantry rifle company, 5-20
sequence of events and times, 5-22 (*illus*)
Defend, 5-31 (*illus*)
graphic scenario, 5-30 (*illus*)
mechanized Infantry company/team, 5-29
sequence of events and times, 5-31 (*illus*)
six day, 5-40 (*illus*)
foothold, moving to the, 2-24 (*illus*)
force-on-targetry, 3-6 through 3-9 (*illus*)
fragmentary order (FRAGO)
FTX Attack, 5-27 (*illus*)
FTX Defend, 5-39 (*illus*)
STX Attack, 5-10 (*illus*)

G

Grenadier Gunnery Trainer, 2-33 through 2-35 (*illus*)

I

Individual and Team Task/Technique Trainer, 2-5 (*illus*)
individual training, 1-11
Infantry rifle
company FTX Attack, 5-20
platoon STX Attack, 5-7

L

leader training, 1-11, D-1
example leader's course POI, D-5
phases, D-3
realistic, off federal facilities, D-5
simulations, D-1
limited visibility training, A-6
Live, Virtual, and Constructive Training Strategy, phases in, D-3 (*illus*)

M

marksmanship training, precision, A-7
mechanized Infantry
company or team FTX Defend, 5-29
platoon STX Defend, 5-12
military operations on urbanized terrain (MOUT), comparison to UO, 1-1 (*illus*)
mission-essential task list, 1-6
movement
across hall, 2-15 (*illus*)
cross-cover formation, 2-16 (*illus*)
foothold, 2-24 (*illus*)
supporting fire position, 2-31 (*illus*)
tactical, 2-14 (*illus*)
multiple room-clearing techniques, 2-9 (*illus*)

N

navigation, A-5

O

obscurity, 2-27 (*illus*)
Offense/Defense Building, Task/Technique Trainer, 2-39
operation order (OPORD)
battalion, 5-32 (*illus*)
company, 3-10 (*illus*)
STX Defend, 5-15 (*illus*)
opposing forces (OPFOR), A-7

P

phases of training, D-3
physical readiness training, A-8
platoon
ODSS STX sectors, 5-4 (*illus*)
sequence of events, 4-5 (*illus*)
tactical scenario (Run), 3-9 (*illus*)
positions for riding on tank, 5-43 (*illus*)
precautions for Infantrymen riding on tank, 5-44 (*illus*)
precision marksmanship training, A-7
preparation
to clear more stairs, 2-19 (*illus*)
to gain foothold, 2-21 (*illus*)
prerequisite training, 1-11
collective task training, 1-13
individual task training, 1-12
program of instruction, D-5 (*illus*)
pyrotechnics requirements for a six-day FTX, 5-40 (*illus*)

R

realistic training, D-5
reserve assault team, 2-30 (*illus*)
risk assessment, 1-6
room-clearing techniques, 2-9 (*illus*)
room, securing of a, 2-29 (*illus*)
Run, 1-10, 3-8 (*illus*)

S

safety
ammunition, A-6
danger areas around tank when firing 120-mm main gun, 5-45 (*illus*)
scenario
Battle Drill 6, varied entry positions, 2-8 (*illus*)
multiple room-clearing techniques, 2-9 (*illus*)
varied, A-8
securing and moving of noncombatants to command post, 2-22 (*illus*)
room, 2-29 (*illus*)
stairs and hallway, 2-18 (*illus*)
wall in an alley, 2-26 (*illus*)
sequence of events and times
FTX
Attack, 5-22 (*illus*)
Defend, 5-31 (*illus*)
platoon, 3-5 (*illus*), 4-5 (*illus*)
Station 3, 2-37 (*illus*)
STX
Attack, 5-9 (*illus*)
Defend, 5-14 (*illus*)

sequence of training, 2-42 (*illus*), 2-47 (*illus*)
SESAMS protection, A-6 (*illus*)
Shoot House, 1-9 (*illus*), 3-1, 3-2 (*illus*)
  training, 3-3
  training and evaluation, 3-15
  training support, 3-14
shoot-through walls and doors, A-9 (*illus*)
simulations training, D-1
situational training exercise (STX)
  Attack, 5-9 (*illus*), 5-10 (*illus*), 5-28 (*illus*)
  Defend, 5-13 (*illus*), 5-14 (*illus*), 5-19 (*illus*)
  sectors, company/team, 5-5 (*illus*)
six-day FTX, 5-40 (*illus*)
squad and platoon task and technique trainer, 2-11 (*illus*)
  scenarios
    Crawl, force-on-targetry, 3-6 (*illus*)
    Run, force on targetry, 3-8 (*illus*)
    Walk, force on force, 3-7 (*illus*)
stability and support operations unit training matrix, C-1
  battalion tasks, C-3 (*illus*)
  brigade tasks, C-2 (*illus*)
  company tasks, C-4 (*illus*)
  platoon/squad tasks, C-5 (*illus*)
Station
  1, Individual and Team Task/Technique Trainer, 2-5
  2, Squad and Platoon Task/Technique Trainer, 2-11
  3, Grenadier Gunnery Trainer, 2-33
  4, Urban Offense/Defense Building, Task/Technique Trainer, 2-39
  5, Underground Trainer, 2-45
supporting
  fire position, moving into the, 2-31 (*illus*)
  tasks for company FTX, 5-20 (*illus*), 5-29 (*illus*)
support requirements, 2-7 (*illus*), 2-13 (*illus*), 2-37 (*illus*), 4-6 (*illus*)

T

T&EOs
  STX Attack, 5-11 (*illus*), 5-28 (*illus*), 5-41 (*illus*)
  STX Defend, 5-19 (*illus*)
tactical
  engagement system, A-5
  exercise without troops (TEWT), 5-6
  movement, 2-14 (*illus*)
tank
  main gun and coax dead space above street level, 5-47 (*illus*)
  riding on, 5-44 (*illus*)
  weapon dead space at street level, 5-47 (*illus*)
target
  requirements, CACTF, 5-41 (*illus*)
  scenario (presentation order) for Station 3, 2-38 (*illus*)
targetry, CACTF, 5-41
tasks, related collective and supporting individual, 3-15 (*illus*)
trainers
  Grenadier Gunnery Trainer, 2-33
  Individual and Team Task/Technique Trainer, 2-5
  Offense/Defense Building, Task/Technique Trainer, 2-39
  Squad and Platoon Task/Technique Trainer, 2-11
  Underground Trainer, 2-45 (*illus*)
training
  audience, Shoot House, 3-3
  Breach Facility, 4-3
  building facade walls, windows, and doors, A-8
training (*continued*)
  collective task, 1-13
  command and staff, 1-11
  concurrent, A-1
  conduct of, 1-10
    Shoot House, 3-4
  demonstrations, A-2
  environments, 1-4
  environment/training phase matrix, 1-3 (*illus*)
  evaluation, Shoot House, 3-15
  evaluators, A-5
  exercise matrix, A-2 (*illus*)
  exercises, A-2
    FTX
      Attack, 5-22 (*illus*)
      Defend, 5-31 (*illus*)
    STX
      Attack, 5-9 (*illus*)
      Defend, 5-14 (*illus*)
  facilities, 1-8
    Breach Facility, 1-9
    combined arms collective, 1-5 (*illus*)
    Shoot House, 1-9
    Urban Assault Course, 1-8
  individual task, 1-12
  information, Shoot House, 3-4
  leader, 1-11, D-1
  limited visibility, A-6
  navigation, A-5
  opposing forces, A-7
  phases, 1-2, D-3
  physical readiness training, A-8
  plan, battalion integrated, 1-6
  precision marksmanship, A-7
  prerequisite, 1-11
  realistic, off federal facilities, D-5
  safety and ammunition considerations, A-6
  sequence and times, 2-7 (*illus*), 2-42 (*illus*), 2-47 (*illus*)
  strategy, 1-2, 1-7 (*illus*)
  support, Shoot House, 3-14
  tactical engagement system, A-5

training (*continued*)
task matrixes, unit, B-1
tips, A-1
use of older or nonstandard facilities, A-9
varied scenarios, A-8

U

Underground Trainer, 2-45 (*illus*)
unit training matrix, B-1
urban
battlespace, 5-42 (*illus*)
offense/defense of building, 2-39 (*illus*)
operations, 1-1
simulations crosswalk, D-2 (*illus*)
training strategy, 1-7 (*illus*)
training support strategy, 1-2 (*illus*)
Urban Assault Course, 1-8 (*illus*), 2-1, 2-2 (*illus*), 2-5
Station 1, Individual and Team Task/Technique Trainer, 2-5
Station 2, Squad and Platoon Task/Technique Trainer, 2-11
Station 3, Grenadier Gunnery Trainer, 2-33
Station 4, Offense/Defense Building, Task/Technique Trainer, 2-39
Station 5, Underground Trainer, 2-45
Urban Operations Training Strategy, Army National Guard 1-7 (*illus*)

V

varied entry positions, 2-8 (*illus*)

W

Walk, 1-10, 3-7 (*illus*)
wall, breaching of a, 4-4 (*illus*)
window, breaching a, 4-3 (*illus*)
wooden facade, 2-34

Made in the USA
Monee, IL
05 November 2026

67807961R00177